Biological Politics

Biological Politics

Feminist and Anti-Feminist Perspectives

JANET SAYERS

TAVISTOCK PUBLICATIONS
LONDON AND NEW YORK

First published in 1982 by
Tavistock Publications Ltd
11 New Fetter Lane, London EC4P 4EE

Published in the USA by
Tavistock Publications
in association with Methuen, Inc.
733 Third Avenue, New York, NY 10017

Typeset by Cotswold Typesetting Ltd
Printed in Great Britain by
Richard Clay (The Chaucer Press) Ltd
Bungay, Suffolk

British Library Cataloguing in Publication Data

Sayers, Janet
Biological politics.
1. Women—Social conditions
2. Sociobiology
I. Title
305.4'2 HQ1154

ISBN 0-422-77870-2
ISBN 0-422-77880-X Pbk

Library of Congress Cataloging in Publication Data

Sayers, Janet.
Biological politics.

Bibliography: p.
Includes index.
1. Feminism. 2. Human biology—Social aspects.
I. Title.
HQ1122.S24 305.4'2 81-16841

ISBN 0-422-77870-2 AACR2
ISBN 0-422-77880-X (pbk.)

Contents

For Nicholas and Daniel

Acknowledgements

I would like to thank my colleagues at the University of Kent for giving me study leave and hence the time to complete this book during the academic year 1979–80. I am also grateful to the staff of Norlin Library, University of Colorado at Boulder, for all the help they gave me in tracking down the source material for this book. Lastly, my thanks to Jan Demarest, Martha Gimenez, David Reason, Sean Sayers, and Shirley Touslon for reading earlier versions of parts or all of the manuscript and for their useful comments on it. I am particularly grateful to Sean Sayers for the time he has devoted to discussing and arguing with me about the ideas in this book, and for the support and encouragement he gave me while writing it.

One

Introduction

This book is about the place of biology in explanations of sexual inequality. I shall be concerned on the one hand with the way in which those opposed to changes in women's social role[1] have sought to appropriate biology for their cause, and on the other with the various ways in which feminists conceptualize the relation between biology and the position of women in society.

Biology has often seemed to lend itself more readily to conservative than to feminist analyses of sexual inequality. Certainly, as I shall try to show in this book, biological accounts of sex differences have often been used to reinforce and maintain traditional sex roles. As Kate Millett puts it, 'patriarchy has a . . . tenacious or powerful hold through its successful habit of passing itself off as nature' (Millett 1971 : 58). It is therefore not surprising to find many feminists being wary of giving any priority to biology in their accounts of sex roles. Since these roles are so easily taken to be 'natural and therefore inevitable', many feminists have indeed 'made every effort only to resort to biological explanations when no cultural or social structural explanations for social phenomena can be found' (Breines, Cerullo, and Stacey 1978 : 48). As a result, although much has been written from the conservative perspective on the influence of biology on women's position in society, there has been a relative dearth of material devoted to this topic from liberal and socialist feminist perspectives.

I shall attempt to go some way towards meeting this lack. The need for an adequate feminist account of the effect of biology on women's social situation is, it seems to me, particularly pressing at this time. The revival, in recent years, of attempts to use biology as grounds for opposing changes in women's social role means that it is now particularly urgent that feminists develop a valid analysis of the relationship between biology and women's destiny, if only to counter the biologically phrased arguments of anti-feminism.

Those opposed to feminism have often taken a 'holier than thou' stance with regard to feminism and biology. They have claimed to accord biology its true place within their accounts of sex roles, and have criticized feminists for neglecting the biological determinants of these roles. This book is organized around a critique of this twofold claim. I shall argue first that the biologically phrased explanations of women's social status advanced by conservative anti-feminists are not, as they claim, grounded in biology. When one examines these supposedly purely biological accounts of sex roles one finds that they are rooted in appeal to social, not biological, considerations. This is true not only of recent biological analyses of sexual divisions in society but also of the analogous biological explanations of these divisions advanced by many of the nineteenth-century critics of feminism. The similarity between earlier and current versions of the thesis that 'biology is woman's destiny' is striking, and I have therefore devoted considerable space to comparing these historically distinct, but theoretically similar defences of sexual inequality.

My concern in the first part of the book has accordingly been to demonstrate the historical and social roots of some of the avowedly purely biological analyses of sexual inequality advanced in support of conservative anti-feminism during the last century. The general thesis that the natural sciences are influenced by, and in turn influence social practice is not, of course, new within the philosophy of science.[2] Nevertheless, despite a few notable exceptions (e.g. Griffiths and Saraga 1979; Hubbard, Henifin, and Fried 1979; Brighton Women and Science Group 1980), there has been a relative lack of interest, especially in England, in demonstrating the current influence of science in reinforcing existing sexual divisions in society. This reflects a general lack of interest on the Left during the post-war period in examining the relation between science and society. There are, however, now signs of an increasing interest in this area (Young 1980; Mackenzie 1981) and I hope that my book will contribute to this development.

The second part of the book addresses the other strand of biological arguments against feminism mentioned above, namely the claim that feminists neglect biology. As well as arguing that the answers to the woman question were to be found in biology, many nineteenth-century writers claimed, for instance, that 'women's physical conformation . . . is utterly ignored by the advocates of sexual equality' (Allan 1869 : 201). Similarly, recent writers have accused feminists of often 'denying the biological component of sex differences' (Lambert 1978 : 114). I shall seek to show that contrary to the claims of their critics, feminists have in fact regularly taken account of biology in explaining the position of women in society.

Feminists are, however, divided as to how biology shapes women's social status. Just as some writers on the general issue of the bearing of nature on society now argue that the 'natural realities are negotiated, conventional and historically alterable' (Young 1977 : 75), so some feminists argue that the influence of biology on women's status is similarly indirect, that it is mediated by the way their biology is interpreted and construed within a given society. I shall term this approach 'social constructionism'. Others, by contrast, argue that biology does affect women directly, that it has endowed women with an essential and particular 'feminine' character from which they have become alienated as a result of living in a male-dominated world. I shall refer to this second position within feminism as 'biological essentialism'.

My own view is that we should forge a third position within feminism, one that takes issue with both biological essentialism and social constructionism. Like many other feminists (e.g. Beechey 1979; Barrett 1980), I believe that biological essentialism wrongly overestimates the biological determinants of women's social status at the expense of neglecting its social and historical determinants. I do not, however, believe that social constructionism is a satisfactory alternative to biological essentialism for it seems to me that in the course of rightly stressing the socio-economic determinants of sexual inequality, social constructionism underestimates its biological roots. Some writers (e.g. Timpanaro 1975) have already provided a critique of the general account provided by social constructionism of the relation between nature and society. My aim is to provide a critique of the specific way in which social constructionism has been developed in opposition to biological determinist and essentialist accounts of sexual inequality. I shall argue that sexual inequality has been determined directly by biological as well as social and historical factors, and I shall seek to demonstrate that this was also the position adopted in early psychoanalytic and marxist accounts of sexual divisions in society.

Notes

1 Some writers (e.g. Angrist 1969; Lopata and Thorne 1978) are unhappy with the use of sex-role terminology. Nevertheless it constitutes a convenient shorthand way of referring to sexual divisions in society and it is in this neutral sense, rather than in the sense of subscribing to a particular (e.g. structural functionalist) analysis of these divisions, that I shall use terms like 'sex role' in this book.
2 Reviews of the history of this viewpoint within the philosophy and sociology of science are provided, for example, by Mulkay (1979) and Filner (1980).

PART I

Biological arguments against feminism

Two

Sexual equality as reproductive hazard

That women bear children and men do not is probably the most important biological difference between them. It is therefore appropriate to begin a discussion of biological arguments on the woman question by considering an argument that takes this biological sex difference as its starting point. In this chapter I shall consider one such argument, one that has been advanced both in the late nineteenth and early twentieth centuries, and more recently today. The gist of this argument is that because of biological factors, sexual equality can be achieved only at the cost of damage to women's reproductive functions. This defence of sexual inequality appears to rest on purely biological premises. I shall try to show, however, that it in fact relies on appeal to social attitudes – to nationalism, to class prejudice, and to sexism.

Education and reproduction

Let us consider first the nineteenth-century version of this argument and how it arose. In 1869 Emily Davies opened Girton College for women and insisted that its students study the same subjects and take the same examinations as men. Newnham Hall was established at Cambridge a year later, and plans were afoot to open the University of Oxford to women; Lady Margaret Hall and Somerville were established at Oxford in 1879 (Hollis 1979).

It was against the background of these social developments that Dr Henry Maudsley, the eminent British psychiatrist, advanced his widely reported biological critique of attempts to allow women equal access to higher education. 'It will,' he said, 'have to be considered whether women can scorn the delights, and live laborious days of intellectual exercise and production, without injury to their functions as the conceivers, mothers, and nurses of children' (Maudsley 1874:471). His own view was that higher education certainly would injure these functions.

Maudsley was not the first to put forward this argument against women's education, though he was perhaps the first in England to address it to the issue of higher education. In 1867 the evolutionary philosopher Herbert Spencer had asserted that 'the deficiency of reproductive power' among upper-class girls

> 'may be resonably attributed to the overtaxing of their brains – an overtaxing which produces a serious reaction on the physique. This diminution of reproductive power is not shown only by the greater frequency of absolute sterility; nor is it shown only in the earlier cessation of childbearing; but it is also shown in the very frequent inability of such women to suckle their infants.' (Spencer 1896:485)

Maudsley, however, differed from Spencer in seeking support for his argument from claims then being made in America about the relation between education, menstruation, and reproduction. In particular, Maudsley cited a book written on this subject the previous year by a former Harvard professor, Dr Edward Clarke.

Like Maudsley, Clarke had written his biological critique of women's higher education not in response to developments within biological theory but in response to social developments – in response to a situation in which more and more colleges were opening their doors to women. The post-Civil War period in America had witnessed an enormous expansion in higher education for women. By 1870 many of the state universities, particularly in the West, had become coeducational. Furthermore, some coeducational and women's colleges were now being established with the express purpose of seeking to emulate the education provided in the prestigious men's colleges of the East. Whereas many of the state colleges channelled women into home economics and education courses and gave them an education inferior to that given to their male students, the newly established private colleges were seriously attempting to provide women with an education as rigorous as that provided for men (Harris 1978). Three such colleges were established in quick succession following the ending of the Civil War. Vassar was founded in 1865, and Smith and Wellesley in 1875. The private men's colleges were also coming under increasing pressure to open their facilities to women and in 1873 Harvard announced that it would be setting examinations to be taken by women under the auspices of the Boston Woman's Educational Association. Successful candidates in these examinations were to be granted accreditation from Harvard (Woody 1929).

Edward Clarke's book, *Sex in Education*, was published in the same year. In it Clarke argued that since educational equality would only be achieved at great cost to women's health – particularly to their reproductive physiology – it was not to be encouraged. The timing of the book's publication, coinciding as it did with the pressure on Harvard to open its facilities to women, gave rise to the not unnatural assumption (e.g. Howe 1874) that it had been inspired not by biological theory, as Clarke claimed, but by the unwillingness of some of Harvard's members, of which Clarke was one, to see women admitted to Harvard. The book clearly filled a contemporary need, for within only a few years of its first publication, it went into seventeen editions (Hall 1905)! Not only was the style of argumentation adopted by Clarke used to inform Maudsley's arguments against women's higher education, but it was also to be used, for at least the next thirty years, by countless writers on both sides of the Atlantic to criticize reforms in the education of women both at the university and at the secondary school level.

As a result of the widespread influence of Clarke's book, the newly established women's colleges in America hired doctors to live on campus to supervise the health of students (Harris 1978). The first president of Bryn Mawr testified to the impact of his book on the cause of women's higher education as follows: 'We did not know when we began whether women's health could stand the strain of college education. We were haunted in those early days, by the clanging chains of that gloomy little specter, Dr Edward H. Clarke's *Sex in Education*' (Thomas 1908 : 69).

Throughout the last decades of the century American writers criticized equal education for girls on the basis of this book. Girls, they said, should not seek to emulate boys intellectually. Their education should instead be tailored to their physiological functions, specifically to the biological requirements of menstruation and reproduction. Nor did these arguments cease with the dawn of the new century. Indeed, the early part of the twentieth century saw a renewed 'reaction against coeducation' and a revival of Clarke's earlier arguments against it. All of those arguments 'that could bear the light of day were unearthed, refurbished, and made to do service' in this cause (Woody 1929 : 280–81). And among them Clarke's argument still seemed serviceable to no less an authority than Dr Stanley Hall, president of Clark University and the acknowledged founder of academic psychology in America.

Drawing on Clarke's arguments and on those of the medical practitioners who had reiterated them during the 1880s and 1890s, Hall asserted that girls of secondary school age should have a different education from boys. The 'data' of these physicians showed, he said, that 'the more scholastic the education of women, the fewer children and the harder, more dangerous, and more dreaded is parturition, and the less the ability to nurse children' (Hall 1905 : 614). During menstruation, said Hall, a girl 'should step reverently aside from her daily routine and let Lord Nature work' (Hall 1905 : 639). Failure to tailor girls' education to what Hall asserted was their physical need for rest during menstruation would, he said, endanger their reproductive powers. Citing Clarke as an authority on the matter, he claimed: 'over-activity of the brain during the critical period of the middle and late teens will interfere with the full development of mammary power and of the functions essential for the full transmission of life generally' (Hall 1906 : 592).

The influence of Clarke's book was not confined to America. As we have seen, his arguments were reiterated by Maudsley in England within a year of their first publication in America. Nor was Maudsley the only doctor to forward this kind of argument in England. In Manchester, for instance, a Dr Thorburn advised, on the basis of Clarke's book, that women should not go to university unless they absolutely required a degree in order to work. Even then, he said, women should be warned that in seeking a university education they were embarking on 'one of the most dangerous occupations of life' (Thorburn 1884 : 18)!

The biological argument

The reason why Clarke's argument seemed so serviceable to those opposed to women's higher education was that it was couched in biological terms and thus appeared to offer a legitimate scientific basis for conservative opposition to equal education. Consider, for instance, the case of the University of Wisconsin. This university, which opened its doors to women in 1860, had for some years opposed the demand that its women students be allowed to pursue the same courses as men (Woody 1929). Just as it seemed to be losing this battle, along came Edward Clarke to provide what seemed to be a watertight biological case for its conservatism on this matter. In 1877 the Board of Regents of the University opposed equal education on the biological and physiological grounds recommended by Clarke. 'Every physiologist,' the Board claimed,

> 'is well aware that at stated times, nature makes a great demand upon the energies of early womanhood and that at these times great caution must be exercised lest injury be done . . . Education is greatly to be desired but it is better that the future matrons of the state should be without a University training than that it should be produced at the fearful expense of ruined health; better that the future mothers of the state should be robust, hearty, healthy women, than that, by over study, they entail upon their descendants the germs of disease.' (Smith-Rosenberg and Rosenberg 1973 : 341–42)

Similarly, we find Sir James Crichton-Browne, a physician, justifying his reactionary attitudes towards the opening of Scottish universities to women with an appeal to Clarke's biological arguments. Overwork, he said, impairs 'the monthly rhythm' and 'the fitness and capacity of the woman to reproduce the species and to bear healthy children' (Crichton-Browne 1892 : 176). And, on these seemingly convincing biological grounds, he condemned the University of St Andrews' recent decision to open its classes to women as 'a downhill step toward confusion and disaster' (Crichton-Browne 1892 : 177).

The basis of these arguments against equal education did appear at first sight to reside in scientific theory. The principle underlying the argument was that of the conservation of energy. This principle had been propounded most clearly for the first time by Helmholtz in 1847. It was a principle which, soon after its first formulation, was used by writers like Spencer both to explain biological phenomena and to develop biological theory (Hofstadter 1955). Scientists working in the Spencerian tradition argued that the principle of conservation of energy as applied to biology meant that energy used up by one organ of the body reduced by the same degree the amount of energy available to other organs of the body. Maudsley, for instance, asserted that 'When Nature spends in one direction, she must economize in another direction' (Maudsley 1874:467). Brain work necessarily detracted, in his view, from the store of energy available for the development of the reproductive system.

Clarke had applied the principle of conservation of energy to the issue of equal education as follows. He assumed that girls have to devote more energy than boys to the development of their reproductive organs during puberty. On the basis of the conservation of energy principle he argued furthermore that energy devoted to academic study would reduce the energy available for the development of the reproductive organs. In common with many physicians of his time he believed that reproductive development in girls took place primarily at the time of menstruation. The fact that menstruation was merely the climax of a menstrual cycle which repeated itself continuously throughout a woman's fertile years was not apparently generally acknowledged until the early years of the twentieth century (Ellis 1929). Clarke reasoned on the basis of his biological knowledge and belief that girls should rest during menstruation in order that their energies be devoted at that time to the development of their reproductive organs; to creating, through the establishment of regular menstruation, 'a delicate and extensive mechanism within the organism – a house within a house, an engine within an engine' (Clarke 1873 : 37). Puberty, in Clarke's view, constituted a critical period for this development. If menstruation was not regularly established then it never would be; reproductive faults occurring in puberty could not, in his view, be rectified later.

Clarke claimed to have deduced his opposition to equal coeducation on the basis of these biological premises. 'Physiology,' he said, not 'ethics', provided the answer to this particular 'problem of women's sphere' (Clarke 1873 : 12). Similarly, Spencer claimed to derive his moral opposition to women's intellectual development on the basis of pure biological theory. He argued that physical and mental development (or 'evolution', as he termed it) normally proceeded less far in women than in men because energy had to be reserved in their case for reproduction:

> 'Whereas, in man, individual evolution continues until the physiological cost of self-maintenance very nearly balances what

> nutrition supplies, in woman, an arrest of individual development takes place while there is yet a considerable margin of nutrition: otherwise there could be no offspring.' (Spencer 1873 : 32)

Women, he said, might be able to equal or even outstrip men intellectually, but since biology had dictated that such intellectual development would be at the cost of their reproductive functions it was to be condemned on biological grounds:

> 'under special discipline, the feminine intellect will yield products higher than the intellects of most men can yield. But we are not to count this as truly feminine if it entails decreased fulfilments of the maternal functions. Only that mental energy is normally feminine which can coexist with the production and nursing of the due number of healthy children.' (Spencer 1873 : 31, n. 1)

Thirty years later Stanley Hall also claimed to deduce his moral objection to women's intellectual development from purely biological premises. Biology, he claimed, showed that the woman who devoted to intellectual development energy 'that was meant for her descendants' was 'the very apotheosis of selfishness from the standpoint of biological ethics' (Hall 1905 : 633).[1]

These writers, then, all sought to legitimate their moral objections to reforms in women's education in terms of biology. Biology, they said, decreed that equal education was morally wrong. But, in fact, as Clarke's detractors pointed out, these objections to women's education were entirely unsound as biological arguments. In the first place, as Elizabeth Garrett Anderson, who was a pioneer in the struggle to gain medical education for women in England, insisted, reproductive organs are formed in girls from birth. They are not newly created in adolescence as Clarke claimed. Furthermore, the development of these organs in puberty does not occur spasmodically only at the time of the menstrual period. As Antoinette Blackwell (a prominent American campaigner for women's rights, as well as for Abolition and Prohibition) pointed out, the development of these organs occurs gradually and continuously throughout adolescence.

Second, even if it was legitimate to apply the conservation of energy principle to biology in the simplistic fashion advocated by Spencer and Clarke – and this was questioned by their critics – (e.g. Blackwell 1875) this principle did not mean that energy devoted to study necessarily diverted energy away from the development of the reproductive organs. Anderson (1874), for instance, also applied the principle of conservation of energy in this matter and like Spencer, suggested that women had 'nutrient' in reserve for child-bearing. But, she argued, these assumptions implied that women could safely use this 'nutrient' for intellectual work until pregnancy made demands on this source of energy.

Lastly, and most importantly, even had the logic of Clarke's application of the conservation of energy principle to biology been watertight as far as his opposition to women's education was concerned, the facts showed the predictions of his theory to be false. The evidence indicated that, far from suffering as the effect of education, women's health appeared to thrive on it. Clarke's own evidence on this point was extremely meagre. Of the seven clinical cases he cited in support of his argument against equal coeducation only one seems to have attended a coeducational college at all. Certainly this woman had a tragic end. She became sick shortly after graduation and died a few years later. Clarke pronounced the case 'an instance of death from over-work' (Clarke 1873 : 103) and the press took it up under the banner 'Educated to death' (Stevenson 1881 : 60). But, in fact, such a conclusion was not warranted by the facts of the case. As one of Clarke's critics quickly pointed out, the girl had fallen ill after she had stopped studying: 'Women sick because they study?' she asked, 'Does it not look a little more as if women were sick because they stopped studying?' (Phelps 1874 : 134). The evidence of the following years suggested that it was 'the want of education' not the 'excess of it' (Stevenson 1881 : 63) that was the main cause of the widely publicized ill-health of middle-class American womanhood. Repeatedly, in the last decades of the nineteenth century, we find middle-class women complaining of nervous and physical disorders, of disorders that they often attributed to 'want of education', and to the idleness that their position in society forced upon them (Ehrenreich and English 1979).

Many of the arguments cited against Clarke by his critics were as subjective and anecdotal as his own. Blackwell, for instance, argued that as she had borne six children since graduation, college education obviously did not harm women's health. Others (e.g. Hayes 1891; Mosso 1892), however, countered Clarke's meagre clinical observations with more substantial empirical arguments based on survey data which showed that the health of graduate women had been largely unaffected by their college experience. Indeed, so sound was the questionnaire study of this issue by Dr Mary Putnam Jacobi, who was then the leading woman physician in America, that Harvard, which had earlier spawned Clarke and his arguments, gave it their 1876 Boylston Essay Prize. Scientifically conducted research has been used, both then and today, to counter arguments against sexual equality. It would be short-sighted to reject science out of hand simply because it has so frequently been called in on the side of those opposed to feminism. In fact, as David Ingleby (1980) points out, the authority of science in the latter respect (i.e. in defending existing social practices) rests on the very fact that its methodology can be, and has been, used to call those practices into question.

The social dimensions of Clarke's argument

Even if one accepted Clarke's biological argument, it did not follow from this argument that women should not be given equal education with men. What he asserted was that women should be given time off study during their menstrual periods. This was not, however, an insurmountable biological obstacle to equal education. There was no reason why colleges should not accommodate this alleged biological need for women by giving them leave from study during their menstrual periods. Since Clarke's argument was in this respect perfectly compatible with the demand that colleges open their doors to women, he sought to bolster it by appeal to a number of non-biological considerations. One consideration, he claimed, was financial. He said that colleges could not possibly afford to accommodate the biological needs of women:

> 'Harvard college could not undertake the task of special and appropriate education, in such a way as to give the two sexes a fair chance . . . without an endowment, additional to its present resources, of from one to two millions of dollars; and it probably would require the larger rather than the smaller sum.' (Clarke 1873:149–50)

Goodness knows why giving women time off for their menstrual periods would be so costly!

Despite the obvious absurdity of this kind of argument it is still being adduced today as grounds for complacency about existing social inequalities between the sexes. Sexual equality, it is said (e.g. Tiger 1970), can only be achieved if institutions make the proper accommodations to the different biological needs of the two sexes, accommodations that these same authors rule out, not on the biological grounds on which they claim to base their objections to feminism, but on the grounds of cost. The 'price . . . required for training and enforcement,' says E. O. Wilson, 'that must be spent to circumvent our innate predispositions' may well be too great to outweigh the advantages of sexual equality' (Wilson 1978 : 148; see also Lambert 1978).

Since Clarke suggested that biology posed women with a choice – a choice between study and reproduction – it was always open to women to argue that they would prefer to study. In this matter, too, his biological argument did not sufficiently prove his case against equal education. As well as appealing to financial sentiment, Clarke also resorted to an appeal to anti-immigrant sentiment to bolster his far from watertight biological argument against reforms in women's higher education.

In 1873, the year in which Clarke's book was published, the number of immigrants to the United States reached an all-time high. Anti-immigrant sentiment was very strong, and had recently led to the tightening of United States immigration policy (Fairchild 1925). Clarke therefore had good reason to believe that he would effectively bolster his argument by appeal to this sentiment. Certainly he repeatedly did appeal to it in prosecuting his allegedly purely 'physiological' argument against women's entry to Harvard. American women could not, be implied, choose to sacrifice their reproductive functions to intellectual work, for if they did so they would run the risk of America being over-run by immigrants. American women, he said, were already much less healthy than immigrant women and, as a result, were less able to bear and rear children. The consequence, he implied, was that the offspring of immigrants would soon vastly outnumber those of American women.

In pursuing this line of attack Clarke cited the arguments previously advanced by his fellow-countryman, Dr Nathan Allen, in his campaign against women's education. The habit of 'too close application to studies' among American women had meant, said Allen, that 'the natural growth and development of the various tissues of the body' had been 'more or less checked' in them (Allen 1870 : 739–40). It was, he suggested, for this reason that whereas English, Scottish, German, Canadian French, and Irish immigrant women had no difficulty nursing their children, American women were often unable to perform this function. A few years later Allen again cited women's education as one cause of the fact that the immigrant population was outnumbering the indigenous American stock on the Eastern seaboard. The New England family, in his view 'the germ of American civilization' (Hall 1905 : 595), was giving birth to fewer and fewer children. This declining birth rate was, he said, only being made good by the constant influx of foreigners and by the high fertility of foreign-born women.

Repeatedly American women were warned that if they continued to pursue the course of equal education they would not only endanger the composition of the future population of their country but would also run the risk that their fellow-countrymen might choose immigrants rather than themselves for wives. Clarke warned that if the number of educated women continued to increase 'the wives who are to be mothers in our republic must be drawn from trans-atlantic homes' (Clarke 1873 : 63). And in 1893, one Cyrus Edson threatened that if American women did not improve in this respect American men would have to have recourse to foreign-born wives (Hall 1905).

Advocates of reforms in women's education did not address this social aspect of Clarke's and his followers' arguments. Indeed they were often as concerned as these anti-feminists lest equality be bought at the price of Americans being outnumbered by immigrants. The way in which some advocates of women's rights have traded on anti-immigrant and racist prejudice in prosecuting their cause is indeed one of the sorrier aspects of the feminist struggle of the last century (Haller and Haller 1974; Magas 1971). In opposing Clarke's argument such feminists retained his nationalistic and anti-immigrant sentiment but rejected his claim that the ill-health of American women was the effect of higher education. Instead they attributed this apparent health problem to other features of the American way of life – to lack of exercise, to the American diet of 'perpetual pie and doughnut' (Anderson 1874 : 592),[2] to the 'laxity of parental authority' in the American family (Anonymous 1874 : 234), and to the concern of American women with their appearance. 'Looks, not books,' said one of the earlier women doctors in America, 'are the murderers of American women' (Stevenson 1881 : 68).

Maudsley had urged women not to study lest they endanger the production and health of future generations of the species. He argued, on grounds of biological theory, that if women pursued the same intellectual goals as men they would inevitably damage their health. Furthermore, on the basis of the Lamarckian principle[3] that women (like men) transmit to their offspring the physical characteristics acquired by them in their lifetime, Maudsley concluded that in injuring their own health through study women would thereby injure the health of their offspring. Women, he said, might choose to pursue careers other than motherhood but, he predicted, such a choice would 'avenge itself upon them and upon their children, if they should ever have any' (Maudsley 1874:482). Female intellectual development was not, in his view, biologically impossible[4] but it was to be morally condemned on biological grounds if, as he argued, it was necessarily bought 'at the price of a puny, enfeebled race' (Maudsley 1874:472). Nor would it do, he said, for intellectual women to choose not to have children: 'Whatever aspirations of an intellectual kind they [women] might have, they cannot be relieved from the performance of these offices [of motherhood] so long as it is thought necessary that mankind should continue on earth' (Maudsley 1874:471).

Maudsley's argument that women's education would result in the deterioration and disappearance of the species was reiterated on both sides of the Atlantic. An American writer, M. A. Hardaker (1882), for instance, said that since intellectual work in women detracted energy from their reproductive functions intellectual equality between the sexes could only be achieved at the cost, within a generation, of the disappearance of the species. And a Dr Withers-Moore, addressing the British Medical Association in 1886, stated that by studying, women drew on their 'capital stock of vital force and energy' thus leaving themselves 'quite inadequate for maternity'. As a result, it was said, 'the human race will have lost those who should have been her sons' (Ehrlich 1975:313).

If, as such writers claimed, their opposition to equal education was motivated solely by biological concern lest such education endanger the survival of the species, why was it that Americans drew no consolation from the high fertility of immigrants, and why was it that both Americans and Englishmen failed to draw any consolation from the high fertility of the working class? The relatively small number of immigrants to England during this period meant that English advocates of Clarke's biological arguments could not appeal, as their American colleagues had, to anti-immigrant sentiment.

They did, however, appeal to nationalistic sentiment. They urged Englishwomen to reproduce for the good of the 'race', and implicitly urged them to outdo the reproductive rates of non-English 'races' (e.g. Thorburn 1884; Pembrey 1913–14). They also appealed to class prejudice to bolster this otherwise very shaky biological argument against women's higher education. A Mr Lawson Tait argued, for instance that: 'To leave only the inferior women to perpetuate the species will do more to deteriorate the human race than all the individual victories at Girton will do to benefit it' (Thorburn 1884:11; see also Allen 1889).

Advocates of this argument in America were also not above appealing to class, as well as to anti-immigrant, prejudice to shore up their opposition to reforms in women's education. Clarke, for instance, asserted that since, in his view, the identical education of the sexes had a particularly 'sterilizing influence' on women its continuation would mean that 'the race will be propagated from its inferior classes. The stream of life that is to flow into the future will come from the collieries, and not from the peerage' (Clarke 1873:139–40). And, over thirty years later, Stanley Hall similarly appealed to class prejudice in this matter. Coeducation, he said, had meant that 'old families are being plowed under and leaders are recruited from the class below'. It would, he warned, therefore lead to 'race suicide' (Hall 1905:605–06).

Clarke's biological arguments have now been dropped from the armoury of anti-feminist polemic. Nevertheless, the relatively low fertility of educated women is still given as a reason why they should be dissuaded from following intellectual careers. And again, one finds appeals being made to class prejudice, prejudice dressed up in the form of intellectual rather than class elitism: 'the dropping out' it is said 'of our most intelligent women from child-bearing and child-rearing must be presumed to result in the subtraction of their assets from those transmitted to the next generation' (Kipnis 1976:117).

Defenders of women's right to equal education in the last century might not have addressed the anti-immigrant bias of the supposedly purely biological arguments advanced against it but they did address the class bias of these arguments. The selectivity with which Clarke advanced his argument against women's participation in middle-class higher education, but not against their participation in working-class occupations, belied his claim to be providing a wholly biological defence of sex discrimination in education. Supporters of the women's rights movement pointed out that since women had to appear for their jobs in the factories and offices regardless of their menstrual periods it was inconsistent to deny middle class women rights on the pretext of their 'temporary disability' during menstruation (Finot 1913:172).

Clarke had anticipated this kind of objection:

> 'Jane in the factory can work more steadily with the loom, than Jane in college with the dictionary . . . the girl who makes the bed can safely work more steadily the whole year through, than her little mistress of sixteen who goes to school.' (Clarke 1873:131; see also Engelmann 1900)

The reason, he said, why the student unlike the working girl needed rest during menstruation, was that menstruation had not been regularly established in her case before she started college. On the other hand, menstruation had, he claimed, been regularly established in those girls who embarked on service work or factory labour. In fact, however, as Clarke's supporters well knew, girls usually entered such paid employment at a much younger age than girls entered college and would therefore have been less likely than them to have started menstruating. They were, however, often willing to grant Clarke license in this matter since it suited their interests to have a respectable, biological-sounding argument to help preserve their monopoly of middle-class educational institutions, an argument that also left intact the benefit they derived from the uninterrupted labour of working-class women.

There was one other social sentiment to which Clarke and his followers appealed in prosecuting their avowedly purely 'physiological' campaign against equal education: namely, sexism. Although they adduced menstruation as a reason for opposing women's entry into traditional, male institutions, they never advanced it as a reason why women should perhaps be relieved periodically from their traditional female work in the home, from work that often involved arduous domestic drudgery throughout every month of their fertile years (Hollingworth 1914).

Furthermore, many of these writers also appealed to male chauvinism and to male fears that sexual equality would threaten their (i.e. men's) job security. Hall, for instance, acknowledged that Clarke's biological argument was 'not very scholarly' (Hall 1905:569) and he sought to strengthen it by appeal to these sexist sentiments.

Coeducation, he said, had resulted in the triumph of women over men,[5] in the 'feminization of the school spirit, discipline, and personnel' (Hall 1905 : 620),[6] and in women ousting men from teaching jobs because they were 'less expensive' (Hall 1906 : 589).[7] Although he claimed to be concerned primarily with women's biological interests, it was, in fact, to the sexist attitudes of men – to their interests in excluding women from competition with them in education and at work – rather than to the biological interests of women that he finally appealed in criticizing coeducation.

Lastly, these critics of equal education all appealed to the sexist prejudice that reproduction is a matter for women not for men. Earlier writers had been quite explicit in equating women, but not men, with their reproductive function. 'Nature,' said one anthropologist, 'has declared, in language which cannot deceive, that women's chief mission is maternity' (Allan 1869 : 201). In 1870, a doctor stated this position even more bluntly. It was, he said 'as if the Almighty, in creating the female sex, had taken the uterus and built up a woman around it' (Smith-Rosenberg and Rosenberg 1973 : 335).

The early twentieth-century critics of women's college education, such as Hall, may not have been so blunt, but they nevertheless appealed to this same social prejudice when they criticized women's (though rarely men's) participation in college education in terms of its effect on their reproductive functions, at least on their fertility. Although male graduates showed less interest in getting married and in having children than did their non-graduate peers, the relatively low fertility of graduate men was rarely cited as a reason for criticizing men's college education. The analogous statistics regarding the low fertility rate of graduate women, however, were constantly being cited as reason why they should not have college education (e.g. Dewey 1886; Gwynn 1898; Hall 1905).

In sum, then, we see that the biological argument adduced by Maudsley, Clarke, and Hall against women's rights in education – their argument that this was not to be tolerated because it endangered women's reproductive functioning – rested on very dubious biological premises. This being the case, the advocates of the argument attempted to render it more plausible by appealing to non-biological considerations. As we have seen, they chose to appeal to financial, anti-immigrant, class, and sexist prejudice to strengthen their biological argument. Although they claimed that their opposition to equal education was derived solely from biology, the plausibility of their argument depended instead on appeal to social sentiment.

Reproductive hazards at work

It might be argued at this point that the claims of Clarke and his followers were obviously ridiculous right from the start and that there is no point in dredging them up again today. The reason that I have given so much attention to their arguments is that, ridiculous though they certainly are, analogous arguments are still being used today against the cause of sexual equality, this time against the cause of equality in industry. It is now being argued, especially in America, that if women work alongside men in certain sectors of industry they will have to do so at the expense of their reproductive functions.

The levels of radiation and of poisonous chemicals in some industrial plants are so high as to be dangerous to the health of the future offspring of employees working in them. On this basis it has been argued that women should either abandon their struggle for equal employment in such jobs and agree to work elsewhere so as not to endanger the health of their future offspring, or should only continue working in these jobs on condition that they get sterilized.

Like the nineteenth-century opponents of equality in education, industry is now attempting to justify the exclusion of women from certain jobs by appealing to what at first sight appear to be purely biological considerations. In fact, however, just as was the case in the nineteenth century, the motivation underlying this recent argument is not anxiety about the welfare of the species so much as anxiety to preserve certain financial interests; and the considerations to which it appeals are social, not biological.

The problem of reproductive hazards in the workplace could be resolved by improving the safety of working conditions. Such a solution has, however, been rejected by the industrial corporations involved on the grounds that such improvement would be too costly and would result in an insupportable reduction in profits (Chavkin 1979). The effect, however, has been that women, rather than the industrial corporations themselves, have had to bear the cost of reproductive hazards at work – either through giving up well paid jobs in occupations entailing such hazards, or through submitting to sterilization.

The emphasis on the reproductive risks to women rather than to men of work in these occupations trades on the social prejudice that reproduction is solely a matter for women. As matter of biological fact, however, many of the substances now adduced as grounds for excluding women from certain sectors of industry are as hazardous to the offspring of men as to those of women. The embryotoxic effects of lead, which are now advanced as reasons why women should be excluded from jobs involving lead-based pigments, are also likely to have an adverse effect on men's future offspring (Chavkin 1979). By trading on the sexist prejudice that biological reproduction is the concern only of women, industrial corporations have been able to deflect some of the pressure that might otherwise have been put on them to make the workplace safe for all their employees, both male and female. Many people, however, refuse to be misled by the claims of industry that it is acting out of pure concern for the biological interests of the species in excluding women from some of its jobs. Some (e.g. East Bay Science for the People 1980) even go so far as to suggest that women should stay on in these jobs and fight alongside men in order to get the working conditions in these jobs made safe for both men and women, and for their offspring. Although this goal is certainly laudable, it seems foolhardy to recommend anyone to continue working in these hazardous conditions.

Just like the analogous biological argument of the last century, this current argument has been applied selectively. The selection of which jobs to scrutinize for reproductive hazards has been determined by social, not biological, considerations. It is those occupations that have, until recently, been all-male preserves that are being inspected for reproductive risks. It is primarily those jobs in industry to which women are only now just beginning to gain admission (just as it was only the traditionally all-male educational institutions to which they were beginning to gain admission in the 1860s and 1870s) that are being scrutinized for reproductive hazards. The American Cyanamid Company plant in West Virginia, for instance, has only recently admitted women to its higher-paying production jobs. It is now arguing, however, that on the grounds of reproductive hazard, women in these jobs should either be sterilized or change jobs. Similarly, the St Joe's Minerals plant in Pennsylvania has hired only very few women to its better paying jobs in the high-lead sections of its operation and it is now transferring many of these workers, again on grounds of reproductive hazard, to other less well paid jobs (Chavkin 1979).

Those jobs that have been more traditionally done by women are not being subjected to the same scrutiny for reproductive hazards. Just as nineteenth-century opponents to women's access to certain jobs and educational institutions on the grounds of their supposed harmful effect on women's health neglected to consider whether work – such as domestic service or sweated needlework – that 'coincided with a woman's natural sphere' (Alexander 1976:63) had similarly harmful effects, so today's opponents to women's access to certain sectors of industry on the grounds of the reproductive hazards involved likewise neglect to consider the possibility that work in traditional female occupations might also entail reproductive risk to the women working in them. Yet many of these occupations do involve real reproductive hazards. Nurses, for instance, suffer as a result of their work a higher incidence of rubella than other members of the population (East Bay Science for the People 1980) – and this is of course a disease renowned for its harmful effects on the embryo. Those nurses who work in operating rooms where they are exposed to waste anaesthetic gases also have a significantly increased rate of miscarriage and of children with congenital abnormalities. Similarly, women who work as electronics assemblers (work that is often assigned to women) are exposed to organic solvents that have been shown to be embryotoxic in several animal studies (Chavkin 1979).[8]

I pointed out above (pp. 20–21) that in advancing their biological arguments for the exclusion of women from secondary schools, writers like Hall also appealed to men's fears that they might be ousted by women from their jobs in these schools. The recent argument that women should be excluded from certain high paying sectors of industry, although couched in apparently purely biological terms, also appeals to the hostility of many men within these sectors of industry to women being employed in these jobs, and to their rejection of sexual equality because of their fears lest women get jobs in these occupations at their expense. In this way industry utilizes sexism – inter-sexual competition within the work force – to bolster its case for having occupational hazards circumvented by excluding women from these jobs, rather than by providing better working conditions for its employees.

It has recently been suggested by Brian Easlea (1981) that men's fears of, and biological arguments against women's emancipation originate primarily from their need to shore up their 'elusive masculinity'. Men's gender identity, says Easlea, remains elusive because, as Nancy Chodorow (1978) suggests, female-dominated childcare has meant that boys are unable to form a secure 'personal identification' in infancy with a primary caregiver of the same sex as themselves. Although this might be one motivating factor for biological arguments against feminism, focus on this psychological factor alone obscures the economic factors that also underlie these arguments.

The arguments of the last hundred years regarding women's entry into higher education and into the work force have also been characterized by some (e.g. Blake 1972) as essentially 'pronatalist'. Women, it is said, were and continue to be admitted to both areas on condition that such entry does not conflict with their reproductive functions. However, the primary consideration in the purportedly biological arguments considered above, first against equal education, and second, today, against equal employment, has been to protect the class interests of an educational élite and of industrial capital respectively. These arguments have certainly appealed to pronatalist sentiments. To characterize them simply as 'pronatalist' is, however, to neglect the selective nature of their pronatalism – the way in which these arguments have repeatedly appealed to nationalistic, class, and sexist sentiment rather than, as they claim, to purely biological considerations regarding species survival.

The appeal to biology to justify the exclusion of women from higher education and from certain highly paid occupations varies as to whether the relationship it postulates between biology and behaviour is one of determinism or of choice. In the latter case, an example of which we have examined in this chapter, biological argumentation has proceeded by requiring that women make the choices, and by moral exhortation to them to make certain types of choice. The perpetrators of such arguments have staunchly resisted the implication of their biological argument that if, as they say, work in certain educational and industrial institutions is hazardous to women's reproductive functions, then these institutions should bear some responsibility for removing these hazards. Where such implications are spelt out, those advancing these apparently purely biological arguments have had to admit to their primary motive in adducing them – namely, that of protecting their economic interest.

Notes

1 A year earlier Roosevelt had similarly condemned the 'viciousness, coldness, shallow-heartedness' of a woman who sought to avoid 'her duty' in respect of reproduction (Gordon 1976 : 142). Women, said such commentators, owed it to society to produce children. One writer even went so far as to assert that women's reproductive functions were the property of society: 'A woman's ovaries belong to the commonwealth,' he said, 'she is simply their custodian' (Van de Warker 1906 : 371).

2 Clarke also criticized American diet in these terms, but had argued that education was the primary cause of the ill-health of American women.

3 The theory of the inheritance of acquired characteristics predated Lamarck but is often attributed to him. It is a theory which was, as we shall see, repeatedly appealed to by nineteenth-century opponents of feminism. This reflects the fact that, even though biologists were often critical of Lamarckianism, they nevertheless regularly used it in elaborating their own theories of evolution. Darwin, for instance, increasingly relied on Lamarck's theory in his later years although he had earlier claimed that he had 'got not a fact or an idea from it' (Darwin 1888, Vol. II : 215). Lamarck's theory only finally collapsed when Weismann's critique of it began to gain currency towards the end of the nineteenth, and at the beginning of the twentieth century. Weismann showed the inadequacy of the evidence for Lamarck's theory. There was, he said, no need to postulate that characters acquired during the individual's lifetime were passed on via the sex cell to his or her offspring; the inheritance of the majority of characters common to individuals of the same species could, he showed, be explained on the assumption of the continuity of the germ plasm across generations.

4 In this respect Maudsley, as well as Clarke, differed from other antifeminists of their day (e.g. Möbius 1901) who, in addition to citing Clarke's arguments as reason why women *should* not aspire to intellectual equality with men, also cited biological facts – in particular, the relatively small size of women's brains – as reason why they *could* not achieve this goal.

5 The educational success of girls evidently touched a sore point for many men in undermining their complacent assumption that they were women's intellectual superiors: coeducation moaned the editors of the *American Naturalist*, had had the unfortunate effect of 'confirming certain doctrinaires in their *a priori* belief in the intellectual equality of the sexes' (Cope and Kingsley 1895 : 826).

6 Hall was not alone in expressing anxiety about the 'feminization' of education. Many of his contemporaries (e.g. Swinburne 1902) voiced similar fears.

7 This kind of complaint against women's entry into the labour force in terms of wages was not, of course, confined to middle-class jobs. At the same time as Hall levelled this complaint against women's entry into the

teaching profession similar complaints were also being levelled against women's entry into working-class occupations (see e.g. Draper and Lipow 1976).

8 Some female-typed occupations are now being surveyed for reproductive hazards but, again, instead of working conditions being improved in these occupations, it is being suggested that women should quit jobs entailing these dangers (i.e. that they should bear the costs of institutional intransigence in this matter). Following the recent radiation leaks at Three Mile Island there has been a move to tighten Federal standards on occupational exposure to radiation. This has been interpreted as meaning not that industry will be forced to eliminate or reduce the radiation hazards in its plants, but that women will be disqualified from working in those plants where radiation levels exceed the Federal standards, from working, for example, as medical technicians in those jobs where exposure to X-rays is such as to be harmful to fetal life (Roeder *et al.* 1980).

Three
Social Darwinism and the woman question

In the last chapter I considered the argument that sexual equality can only be achieved at the cost of harming women's reproductive functions. I shall now consider some of those biological arguments about sex roles that have been couched in evolutionist terms. Such arguments have become increasingly influential in the last few years and I shall devote the next three chapters to a consideration of them.

I shall be concerned, in this chapter, to outline some of the historical precursors of modern evolutionist accounts of sex roles. It has recently been claimed that these accounts, and the application of evolutionary theory to social behaviour in general, constitutes a 'new' departure within the behavioural sciences for which the term 'sociobiology' has been coined. I shall argue that this is by no means a novel use of evolutionary theory. This biological theory was repeatedly used in the late nineteenth and early twentieth centuries to confront the social questions, including the woman question, of the day. I shall also show that, just as some feminists today accept the premise of sociobiology that social behaviour is ultimately reducible to evolutionary biology, so some of the earlier feminists also accepted this premise as it was put forward then by writers like Herbert Spencer. Lastly, I shall argue that in accepting this premise these feminists, like the social Darwinist anti-feminists with whom they took issue, wrongly neglected the fact that human behaviour, though importantly affected by biology, is also affected by society.

Modern evolutionist arguments against feminism

The recent resurgence of feminism has been countered by some behavioural scientists with the claim that traditional sex roles are the effect of evolution, and that attempts to change them are therefore futile and morally wrong. E. O. Wilson, for instance, writes in the following terms of the futility of the women's movement. Evolution, he says, working through genetic mechanisms, has determined contemporary sex roles. As a result, he claims, they are relatively immutable:

> 'In hunter-gatherer societies, men hunt and women stay at home. This strong bias persists in most agricultural societies and on that ground alone, appears to have a genetic origin . . . the genetic basis is intense enough to cause a substantial division of labor even in the most free and most egalitarian of future societies.' (Wilson 1975b: 48, 50)

Similarly a number of psychologists have also asserted that contemporary sex roles are the effect of evolutionarily determined psychological sex differences. Jeffrey Gray and Anthony Buffery assert, for instance, that the traditional division of labour between the sexes – 'the mother-infant nuclear unit and the generally greater role played by males in competitive social interaction' – is the effect of psychological sex differences evolved during our early primate past. It is, they say, therefore of 'essentially biological origin', and is 'specified in the gene pool of our species' (Gray and Buffery 1971: 107). John Crook likewise claims that existing psychological sex differences originated 'early in human history', that the development then of a more carnivorous diet and of tools for hunting resulted in women's 'psychosocial attributes' becoming 'more suited to homemaking and support, man's to outward exploration, hunting and control over events' (Crook 1972: 248, 249). Adducing the same kind of argument, Corinne Hutt concludes that contemporary sex roles are therefore the 'evolutionary heritage of modern man' (Hutt 1972: 107).

On the basis of such claims these writers go on to condemn some of the recent demands of the women's movement as not only futile but also morally wrong. Crook, for instance, suggests that evolution has dictated that the relations between the sexes be essentially harmonious, and has criticized the women's movement for seeking to disrupt this harmony:

> 'Those concerned with Women's Liberation would be wise to ponder the biological and psychological complementarity of the two sexes and their deep emotional needs for partnership as a counter to the notion of a poorly defined "equality".' (Crook 1972:275)

Hutt also insists on the biologically ordained complementarity of psychological sex differences and of the sex roles to which she maintains they give rise. On this basis she criticizes feminists, who she represents as seeking to compete with men rather than cooperate with them. She implies that women's primary social role not only is, but also should be, that of motherhood and that women should not strive for 'masculine goals' nor seek to enter 'the competitive, assertive spheres of the male' (Hutt 1972:138).

I shall argue in subsequent chapters that the biological premises adopted by these writers do not warrant their conclusions as to the futility or the wrong-headedness of the contemporary women's movement. I shall be concerned here, however, to address another claim put forward by these writers, and in particular by E. O. Wilson (1975a), namely, that such arguments, as they are now being used to address social issues like that of sexual inequality, represent a 'new synthesis' of social and biological theory. This synthesis is, I shall argue, not at all new. During the late nineteenth and early twentieth centuries similar syntheses of social and biological theory were repeatedly advanced in argument about such social questions. In using the scientific authority of evolutionist theory to support their conservative opinions on political questions writers like Wilson are, in fact, following in the long-established tradition of social Darwinism.

Darwinian theory and nineteenth-century anti-feminism

Darwin himself, in *The Origin of Species*, resisted spelling out how his theory of evolution might apply to humanity. He simply anticipated that as a result of his theory, 'Light will be thrown on the origin of man and his history' (Darwin 1968:458).[1] Later, in 1871, he did develop his earlier ideas about sexual selection[2] and used these ideas to explain the evolution of physical and behavioural differences between the sexes.

Although E. O. Wilson now claims some originality in applying

Darwin's theory of evolution to social issues, this theory was, in fact, seized on immediately it was first formulated and used to address the political questions of the day. Within a year of the publication of *The Descent of Man*, the English economist and journalist, Walter Bagehot (1872), attempted to use the principle of natural selection to justify his own conservative political views (Cowles 1936). Like the sociobiologists of today (see Chapter 4 below), Bagehot took certain liberties with Darwin's theory in order to make it more serviceable to his political arguments.

Consider, for instance, Bagehot's evolutionist account of sex roles. Darwin claimed that behavioural sex differences were primarily the effect of sexual rather than of natural selection. They had been selected, he suggested, as a result of competition for sexual partners (i.e. of sexual selection), not as a result of the general struggle for existence (i.e. of natural selection). If, he said, such differences had 'been accumulated through natural selection', that is 'in relation to the ordinary habits of life' this would imply 'that the two sexes follow different habits in their struggles for existence, which is a rare circumstance with the higher animals' (Darwin 1896: 241).

Bagehot, by contrast, asserted that natural (not sexual) selection was responsible for the behavioural and physical differences between the sexes that Darwin had described. These differences, he said, had evolved not through competition for mates but through the struggle for existence in which the sexes played different roles – the male that of defender, and the female that of nurse. On the basis of this departure from Darwin's theory Bagehot went on to assert that existing human sex roles had been given, and were fixed by evolutionary biology. 'Each sex,' he said, 'fulfils the tasks for which it is especially adapted by Nature' (Bagehot 1879: 208). Therefore, he argued, the women's rights movement was futile. In seeking for equality in sex roles it was conducting 'a struggle against Nature; a war undertaken to reverse the very conditions under which not man alone, but all mammalian species have reached their present development' (Bagehot 1879: 208).

Falsely equating, as did many of the social Darwinists, human *history* with human *evolution*, Bagehot asserted that since sex roles had become more differentiated in the course of history they must on that account alone be the product of evolution just like the behavioural sex differences observed in the animal kingdom. In his view, the women's movement, by attempting to reduce these role differences, was thereby seeking to alter not merely the course of history but the whole course of biological evolution as well. The movement was, therefore, in his view, 'palpably retrograde'. It was 'an attempt to rear, by a process of "unnatural selection", a race of monstrosities – hostile alike to men, to normal women, to human society, and to the future development of our race' (Bagehot 1879 : 213).

Others, like Bagehot, also appealed directly to Darwin's theory in seeking to bolster their opposition to the women's rights movement. And, like Bagehot, they also refurbished it to make it more serviceable to their cause. (One such writer – Cope (1888) – even went so far as to reinterpret Darwin as having stated that natural selection only applied to men, not to women!) However most of those writers of the period who used evolutionist arguments to address the issue of women's rights utilized Spencer's not Darwin's theory for this purpose; a theory which, though it acquired much of its respectability from Darwin's theory, was derived in large measure independently of it (Peel 1972).

Spencerian anti-feminism

Just as E. O. Wilson's use of evolutionism to address many of today's social issues has met with widespread acclaim,[3] so Spencer's similar use of evolutionism in the nineteenth century was also lauded by his contemporaries. The appeal in both cases has been that, despite the flaws in their arguments, they both seem to provide respectable biological justification for moderate, and even extreme[4] conservative opinion on a number of social issues. Since Spencer's view had a great influence on nineteenth- and early twentieth-century arguments about the woman question, I shall outline them in some detail.

In his earlier days Spencer had supported women's rights. He argued in 1850, for instance, that the liberal doctrine of equal rights – the 'law of equal freedom' – implied that women should, in all consistency, be given equal rights with men (Spencer 1884 : 173). By 1873, however, he was arguing against women's rights, at least their right to the vote, on the grounds that biology had made them more charitable than men and thus more likely to interfere with the natural course of social progress by giving help to society's weakest members, to those who in the natural struggle for existence should not survive.[5]

In the years between 1850 and 1873 Spencer had gradually developed a biological, an organic, account of social progress. Already in his *Social Statics* of 1850 he had used the Lamarckian principle of the inheritance of acquired characteristics to deduce that human progress was inevitable. Since he said, human 'imperfection' represents 'unfitness to the conditions of existence' (Spencer 1884 : 79), such imperfections would die out through disuse. On the other hand, those faculties which were adapted to existing conditions would, he argued, increase through use and be passed on to succeeding generations.

Two years later he further elaborated his biological and optimistic account of social progress ironically by appeal to Thomas Malthus's extremely gloomy doctrines on population. Malthus had developed his theory as a critique of the optimistic social viewpoints of Godwin and Condorcet. He had argued that the 'strong and constantly operating check on population from the difficulty of subsistence' necessarily results in 'misery and vice' and is 'decisive against the possible existence of a society, all the members of which should live in ease, happiness, and comparative leisure' (Malthus 1960 : 9–10). Spencer, by contrast, used Malthus's theory of population to argue that 'pressure of population upon the means of subsistence' (Spencer 1852 : 498) would eventually lead to the perfect adaptation of humanity to its conditions of existence, and to the 'survival of the fittest' (Spencer 1852 : 498), since only those who advanced in terms of skill, intelligence, and self-control would win out in the struggle resulting from this pressure.

Spencer had referred to Malthus's social theory as a theory of 'animal fertility'. Seven years later Darwin also asserted that Malthus's law applied to the whole living world:

> 'as more individuals are produced than can possibly survive, there must in every case be a struggle for existence, either one individual with another of the same species, or with the individuals of distinct species, or with the physical conditions of life. It is the doctrine of Malthus applied with manifold force to the whole animal and vegetable kingdoms.' (Darwin 1968 : 117)

And this social 'doctrine' was to constitute a cornerstone of his theory – the theory that evolution is the effect of natural selection or, to use Spencer's term, of the 'survival of the fittest'.

Darwin (e.g. 1968 : 348) rejected the idea that evolution

guaranteed social progress. Spencer, on the other hand, believed that it did. Furthermore, he claimed that evolution also dictated the character of this progress. 'Organic progress,' he said, 'consists in a change from the homogeneous to the heterogeneous.' Since this 'law of organic progress' was, in his view, the 'law of all progress' (Spencer 1857 : 39–40), so biology had therefore dictated that social progress would be marked by an 'advance . . . towards greater heterogeneity' (Spencer 1857 : 48). In the further development of his theory Spencer was to conclude that this 'organic law' as applied to social progress entailed a proliferation in, and a 'spontaneous evolution' of, the division of labour in society. That this evolution was spontaneous could, he said, be demonstrated by the fact that specialization in industrial organization, for instance, (e.g. Yorkshire's specialization in woollen and Lancashire's specialization in cotton manufacture) had proceeded 'in spite of legislative hindrances' (Spencer 1860 : 54).

Nearly a century previously, in 1776, Adam Smith had eulogized the division of labour as one of the hallmarks of social progress: 'The greatest improvement in the productive powers of labour, and the greater part of the skill, dexterity, and judgment with which it is anywhere directed, or applied, seem to have been the effects of the division of labour' (Smith 1937 : 3). But whereas Smith had recognized that this development was the effect of economic factors – of 'the extent of the market' (Smith 1937 : 17) – Spencer now claimed that it reflected the workings of biology, or 'organic law', of a law that the women's rights movement, among others, wrongly sought to flout.

Like Bagehot, he claimed that the division of labour between the sexes within his own society was the result of evolution. Applying to the woman question his view that Lamarckianism and Malthusianism implied the increasing adaptation of humans to their conditions of life, he reasoned that biology must have fitted the sexes to their different social functions, and that, in turn, existing sex roles were therefore biologically prescribed. Like the writers of the 1970s cited above, he argued that existing psychological sex differences had been evolved to fit the sexes for their respective roles in the sexual division of labour. He premised his argument to this effect on the claim:

'That men and women are mentally alike, is as untrue as that they are alike bodily . . . To suppose that along with the unlikeness between their parental activities there do not go unlikeness of mental facilities, is to suppose that here alone in all Nature there is no adjustment of special powers to special functions.' (Spencer 1873:31)

Having thus assumed what he sought to prove, Spencer went on to characterize the differences between the sexes entirely in terms of this assumption, in terms of the premise that these differences were perfectly adapted to the existing social roles of the two sexes.

Earlier, in 1850, he had criticized the use of the existence of particular sexual inequalities as justification of those inequalities:

'As the usages of mankind vary so much, let us hear how it is to be shown that the sphere *we* assign her [woman] is the true one – that the limits *we* have set to female activity are just the proper limits. Let us hear why on this one point of our social polity we are exactly right, whilst we are wrong on so many others.' (Spencer 1884:169. Spencer's emphasis)

In 1850 he argued that the existence of sexual inequality in no way justified it. Now, in 1873, he argued that since men and women performed different roles in society this inequality must be justified on that basis alone. Nature, he said, must have fitted the two sexes for their current social roles – women as childrearers, men as workers in the public domain. Women, he said, must have been fitted by biology with a 'parental instinct' well adapted to childcare since this was their role in society. He thus attempted, albeit in circular fashion, to justify existing sex roles as the effect of nature, and he ended his article by reiterating its premise – a premise that now featured as its conclusion – namely the justification of 'the *a priori* inference that fitness for their respective parental functions implies mental differences between the sexes' (Spencer 1873:35).

In Spencer's view biology had not only ordained the fitness of men and women for their different social roles, it had ordained the very division of these roles in the first place; the sexual division of labour, like all divisions of labour, was, he said, a product of the organic law of progress, the law of increasing heterogeneity. Social progress had in his view been inevitably marked by increasing differentiation of sex roles: 'On ascending to societies partially or wholly settled, and a little more complex, we begin to find considerable diversities in the division of labour between the sexes' (Spencer 1898 : 730–31).

He also adduced the biological principle of the 'survival of the fittest' to justify his claim that the sexual division of labour that characterized the middle-class family of his day represented the height of evolutionary achievement. Societies marked by such a division would, he said, necessarily triumph over societies with greater equality in sex roles:

> 'the societies in which these available males undertake the harder labours, and so, relieving the females from undue physical tax, enable them to produce more and better offspring, will, other things equal, gain in the struggle for existence with societies in which the women are not thus relieved.' (Spencer 1898 : 743)

Although Spencer (1898 : 767) maintained that there was room for further developments in the 'emancipation of women' he also maintained that the claims of the feminists had, in other directions, been pushed too far. Evolution, he said, had led to the present division of labour between the sexes: 'up from the lowest savagery, civilisation has, among other results, caused an increasing exemption of women from bread-winning labour . . . in the highest societies they have become most restricted to domestic duties and the rearing of children' (Spencer 1898 : 768). Therefore, he said, it would be entirely 'mischievous' to educate women so as to fit them for 'business and professions' (Spencer 1898 : 769).

Spencer's criticism of the women's movement was to be constantly reiterated in the following years. Some argued that since as Spencer had claimed, an increasing differentiation of the roles of the two sexes was guaranteed by biological evolution there was nothing to be feared from the 'Women's Rights Viragoes', for, it was said, 'They might as well try to convince women to wear beards, or men, crinolines (Finck 1887 : 174, 176). Other writers also cited evolutionist theory as reason why the women's rights movement might be viewed with equanimity; for, said one such writer, 'You may drive out Nature with a pitchfork, but she will inevitably return' (Bloch 1909 : 70).

Other Spencerians did not feel so happy about dismissing the women's movement as ultimately innocuous. If the demands of the feminists were granted, said one Spencerian, the course of evolutionary progress would be entirely disrupted. Female suffrage, he said, was 'distinctly, emphatically, and essentially retrograde in every particular' and was therefore to be vigorously opposed (Weir 1895:818). Similarly others, who were more pessimistic than Spencer about the inevitability of social progress, claimed that coeducation, equal employment, and the suffrage would speed on 'the retrogressive period of [human] evolution' and were, for this reason, to be entirely rejected (Hyatt 1897:91).

As Havelock Ellis pointed out, however, if, as these writers claimed, evolution had dictated women's destiny it was also likely to be robust enough to withstand changes in that destiny:

> 'We may preserve an attitude of entire equanimity in the face of social readjustment. Such readjustment is either the outcome of wholesome natural instinct, in which case the social structure will be strengthened and broadened, or it is not; and if not, it is unlikely to become organically ingrained in the species.' (Klein 1971:43)

Nevertheless, the use of Spencerian evolutionism to condemn the women's movement as threatening evolutionary progress continued unabated in the succeeding decades (see, for example, Ferrero 1894; Sedgwick 1901).

And, just as Clarke (see Chapter 2 above) had appealed to nationalistic and class prejudice to advance his supposedly purely biological argument against the feminist cause, so the Spencerians also appealed to the same social sentiments in advancing their arguments about the woman question – sentiments which they dressed up in evolutionist terms. Surely, it was asked, middle-class women did not want to become like working-class or non-European women? But that was exactly what they would do, it was implied, if they persisted in demanding sexual equality, since equality was something that only occurred in these 'less evolved' sectors of society. The 'equality of the sexes', said one such writer,

> 'occurs merely . . . in the lower classes of society. On the other hand, the pre-eminence of the male as compared with the female marks a higher stage of evolution. It occurs in the highest species and races, in the prime of life, and in the superior strata of society.' (Fernseed 1881:744)

Like Clarke's argument, this one also took its character not from biology, as was claimed of it, but from the ethnocentric and class prejudiced attitudes of its advocates. And, like all social Darwinist arguments, it relied on the false equation of history with evolution. It attributed the supremacy of middle-class Europeans to biology and to evolutionary progress, when in fact it had been the result of historical and economic processes, processes that had led to the exploitation of the working class at home and to the colonization of other races abroad.

Geddes and Thomson on The Evolution of Sex

Before considering how feminists at the time responded to such arguments I shall outline one last nineteenth-century theory about the evolution of sex differences. Despite its well justified obscurity today this theory was in its time almost as influential as Spencer's in informing contemporary discussion of the woman question. It was developed as a response to criticisms then being made of Darwin's theory of evolution.

Paradoxically, at the same time as social Darwinist ideas, in the form of Spencerian doctrine, were reaching the height of their influence on social thought, Darwin's reputation in biology was reaching its lowest ebb. The decline of Darwin's ideas in biology had resulted in part from new evidence suggesting that the word was younger than was compatible with Darwin's time-scale for evolutionary development, and from Darwin's failure to explain how variation in offspring – the material basis for natural selection – occurred (Burrow 1968).[6]

The breach in evolutionary theory threatened by this development was filled, as far as the evolutionist account of human sex differences was concerned, by the Scottish naturalist turned sociologist Patrick Geddes, and by his pupil J. Arthur Thomson. Darwin, as we have seen, had attributed these differences primarily to sexual selection, although as he turned from explaining secondary sexual characters in infra-human species to explaining them in humans, he began to accord natural selection a more equal place with sexual selection as a cause of these characters:

> 'With social animals the young males have to pass through many a contest before they win a female, and the older males have to retain their females by renewed battles. They have, also, in the case of mankind, to defend their females, as well as their young, from enemies of all kinds, and to hunt, for their joint subsistence. But to avoid enemies or to attack them with success, to capture wild

animals, and to fashion weapons, requires the aid of the higher mental faculties, namely, observation, reason, invention, or imagination. These various faculties will thus have been continually put to the test and selected during manhood . . . the higher power of the imagination and reason . . . will have been developed in man, *partly through sexual selection,* – that is, through the contest of rival males, and *partly through natural selection,* – that is, from success in the general struggle for life; and as in both cases the struggle will have been during maturity, the characters gained will have been transmitted more fully to the male than to the female offspring.' (Darwin 1896 : 564–65. My emphasis)

Geddes and Thomson, writing in 1889 when Darwinism was suffering a decline in the estimation of biologists, echoed the opinion of many of their contemporaries in asserting that 'the shoulders of natural selection' had 'been overburdened' by Darwin. Natural selection, in their view, played only a limited role in the development of sex differences and they dismissed sexual selection as a teleological notion, a concept which they said constituted only a special case of natural selection and which failed to explain the origin, as opposed to the elaboration, of sexual differences.

Weissmann had argued that inherited characters were transmitted in the germ plasm alone, and that the plasm remained constant from one generation to the next, and was unaffected by changes occurring in the organism as the effect of experience. This theory, which began to become known in England in the 1880s, presented an apparent difficulty to Darwinians. Such scientists, ignorant of Mendel's discoveries, of modern genetics, and of recent discoveries about the causes of genetic mutation, could not begin to explain how it was that if Weissmann's theory was correct and the germ plasm remained unchanged across generations, biological mutations and hence evolutionary change occurred. While some evolutionists (e.g. Romanes 1893) simply rejected Weissmann's theory out of hand on this account. Geddes and Thomson adopted Weissmann's theory but suggested that the 'general protoplasm' surrounding the germ plasm (i.e. the cell nucleus) might also be important in fertilization. Such a suggestion not only had the advantage of explaining how variations might take place across generations via the effect of the environment on cell protoplasm; it also provided these writers with the means of giving a new biological gloss on human sex roles.

The metabolism of this cell protoplasm varied, they suggested, with changes in the environment. Where there was a plentiful supply of food this metabolism favoured the formation of female embryos, whereas in conditions of relative scarcity of food resources the metabolism of this protoplasm favoured the formation of male embryos. In like fashion, they suggested, the metabolism of the female sex cell, and of all female body cells in turn, involved the constructive transformation of incoming food supplies (a process they referred to as 'anabolism'). On the other hand, the metabolism of male cells consisted primarily of the destruction of food reserves, of the breaking down of cell products (a process they referred to as 'katabolism'). In their view the psychological differences between the sexes reflected these basic differences in cell metabolism, and were not, as Darwin had suggested, the effects of sexual and natural selection. Men's katabolic metabolism, they said, had made them 'more active, energetic, eager, passionate, and variable', and had resulted in their having more 'scientific insight' and a 'stronger grasp of generalities'. Women's anabolic metabolism on the other hand had, they claimed, made them 'more passive, conservative, sluggish, and stable' (Geddes and Thomson 1890: 249–50)![7]

Whereas Darwin had regarded some, at least, of these psychological sex differences as modifiable by education,[8] Geddes and Thomson were adamant that since these differences were an expression of differences in cell metabolism, differences that prevailed in all species marked by sexual dimorphism, they could not be modified. 'What was decided among the prehistoric Protozoa', they declared, 'can not be annulled by Act of Parliament' (Geddes and Thomson 1890: 247). And with this slogan they provided a new rallying cry for those who wished to render their anti-feminism biologically respectable.

Just as recent writers have proclaimed that the psychological and hence social complementarity of the two sexes is given by biological evolution, so Geddes and Thomson also maintained that evolution had dictated that human sex roles should be complementary. Since, they said, the metabolic processes of the two sexes complemented each other, biology had thereby ordained that the roles of the two sexes should likewise complement each other. That the husband in some societies lay idle after his return from hunting while his wife was 'heavy-laden' and 'toils and moils without complaint or cease' was, they said, fully justified by biology. Their argument ran as follows. Male metabolism involves 'extreme bursts of exertion'; thus men have to utilize 'every opportunity of repose'. The division of labour in which men rest while women work was therefore, they claimed, 'the best, the most moral, and the most kindly, attainable under the circumstances' (Geddes and Thomson 1890:248). Not content with making such outrageous attempts to justify the exploitation of women in other societies, they went on to argue that biology had similarly ordained prevailing inequalities between the sexes in their own society. Biology, they said, had ordained that it would be wrong to raise women's wages, or indeed to allow them entry into 'the competitive industrial struggle' (Geddes and Thomson 1890:247). In general, they maintained, evolutionary progress lay in the 'complex and sympathetic co-operation between the differentiated sexes' (Geddes and Thomson 1890:249). Morality required adherence to the dictates of biology: 'The social order will clear itself, as it comes more in touch with biology' (Geddes and Thomson 1890:249). Compliance with the ideals of masculinity and femininity was, in its turn, they implied, not merely biologically determined but also morally essential. Masculinity and femininity were given by biology, and this meant that it was the moral duty of both men and women to conform to these supposedly biologically given ideals for their respective sexes.

Despite the absurdity of many of their claims, Geddes's and Thomson's theory was to be used, like Spencer's ethnocentric theory before it, in order to bolster reactionary responses to women's issues. Like both Spencer in the nineteenth century and Wilson today, Geddes and Thomson enjoyed immense publishing success. Their book, *The Evolution of Sex*, was issued in America within a year of its first publication in England. It was also soon issued in French translation. As a result we find its arguments appearing in many places in debates about women's proper role in society (e.g. Thomas 1897; Fouillée 1895). Indeed, so serviceable did Geddes's and Thomson's arguments seem to the anti-feminist cause that they were still in circulation sixty years later when de Beauvoir criticized them in *The Second Sex*.

In the late nineteenth century the evolutionary theory of writers like Geddes and Thomson was used to criticize coeducation in the university, and the entry of women into occupational life. The areas singled out for scrutiny by such social Darwinists were thus different from those singled out by present-day anti-feminists. Today, the focus is more often on sexual equality at work and on sex roles within the family. As a further consequence of the gains made by women over the last century, current claims regarding the biological determinants of sexual inequality are usually couched in more moderate terms than those of their nineteenth-century forebears. Nevertheless, those using Geddes's and Thomson's theory to criticize the women's rights movement of their day often used arguments essentially similar to the modern evolutionary theory used to criticize today's women's movement. Just as the psychologist, John Crook, now uses evolutionary theory as a basis for criticizing feminists for, as he views it, seeking to disrupt the naturally-given harmony between the sexes, so the physician, Sir James Crichton-Browne, for instance used the evolutionist theory of his day (that of Geddes and Thomson) to argue that biology had ordained a 'reciprocal dependence and higher harmony' between the sexes, one to which those demanding sexual equality were, he said, blind (Crichton-Browne 1892 : 177–78). Similarly, just as the psychologist Corinne Hutt today uses modern evolutionary theory to argue that women should place 'value and emphasis on their particular talents and skills' and at the same time dismisses such skills (e.g. their verbal skills) as mere 'fluency' (Hutt 1972 : 138, 96), so an earlier psychologist and philosopher, G. T. W. Patrick, used Geddes's and Thomson's evolutionist theory to argue both that women's talents were 'too sacred to be jostled in the struggle for existence' and at the same time characterized women as evolutionarily 'retarded' (Patrick 1895 : 225, 219).

Feminists for social Darwinism

Many supporters of the women's movement rightly dismissed the arguments of Geddes and Thomson as pure cant. Their 'theory of the essentially conservative nature of woman's intelligence' was, said one writer, an obvious 'fiction of the male intelligence, maintained in order to keep this inconvenient radicalism of woman in check' (J.D. 1894 : 407).[9] Other feminist sympathizers, however, agreed with the main tenet of social Darwinism – namely, that social behaviour should be guided by evolutionary theory. They simply disagreed with the conclusions that writers like Geddes and Thomson and Spencer drew from this theory as far as women's rights were concerned.

Antoinette Blackwell, for instance, who we have come across before as a staunch critic of Edward Clarke's biological arguments against women's university education, agreed with many of

Spencer's arguments about women's proper role in society. And, like Spencer, she believed that evolutionary theory provided an adequate basis on which to decide that role. The division of labour between the sexes in which women looked after the children while men protected them was, in her opinion, just as Spencer had said it was, a division that was governed by the laws of evolutionary progress, and one that should not therefore be questioned.

Her main quarrel with Spencer and with Darwin lay with their failure to give equal value to masculinity and femininity. Spencer had asserted that evolution rendered women mentally inferior to men (see Chapter 2, above), and had not attributed to them any compensating psychological virtues. Darwin had written in a similarly disparaging vein of women's mental faculties:

> 'It is generally admitted that with woman the powers of intuition, of rapid perception, and perhaps of imitation, are more strongly marked than in man; but some, at least, of these faculties are characteristic of the lower races, and therefore of a past and lower state of civilisation.' (Darwin 1896 : 563)

Blackwell, by contrast, asserted that biology dictated that masculine and feminine traits be entirely equal in value even though very different in character especially once one reached the evolutionarily advanced stage of humanity. 'Disequilibrium', she said, would have resulted had these traits not been equal in value, as she argued that children inherit equally from their mothers and fathers and their characters would have been unbalanced had their mothers' and fathers' traits been unequal in value. 'Natural selection', she therefore concluded, 'must tend to maintain equivalence in the two sexes of every species, and to carry forward all evolution on two mutually adapted lines' (Blackwell 1875 : 33).

It was on the basis of these supposedly valid evolutionary considerations that she opposed the existing social restrictions on women. She roundly criticized the 'regimen of sofas' and the 'mental torpor' inflicted on women of her class on the grounds that this had resulted in their feminine faculties not being developed to equal the masculine faculties of men. Such an imbalance between masculinity and femininity meant that the future offspring of the species would be deprived of the endowments they should have inherited from their parents. The 'human race,' she said, had, as a result of its 'arrogant repression' of women, been guilty of 'forever retarding its own advancement, because it could not recognize and promote a genuine, broad and healthful equilibrium of the sexes' (Blackwell 1875:117–18). Nature might have compensated in some measure for the effects of making woman into 'a doll or a drudge' but, she went on, 'It must be sheer folly to believe that the offspring of such a one will not be defrauded of the increase which should revert to them from the exercise of parental talent' (Blackwell 1875:118). Thus, unlike the anti-feminist social Darwinists considered above, Blackwell argued that it was the repression, not the liberation, of women that was the more serious obstacle to human evolution.

Other feminists likewise used social evolutionist arguments to support their cause. In 1898, for instance, Charlotte Perkins Gilman (the then leading intellectual of the American women's movement) defended the existing division of labour between the sexes in the same evolutionist and ethnocentric terms as Spencer had used before her: this division had, she said, led to 'enormous racial gain' (Gilman 1966:135). But she went on to argue, contrary to Spencer, that further social progress now depended on the granting of social equality to women. Existing social restrictions on women, she said, deprived them of the experience necessary to the development of social virtue. As a result, she claimed, women had less well developed moral qualities than men. And this, in turn, meant in her view that women were producing children with a 'hybrid' moral character made up of the moral deficiencies of the mother alongside the moral virtues of the father. Evolutionary progress, she concluded, was thus being retarded by the existing social inequalities between the sexes, and would continue to be so as long as women remained 'tied to the starting-post, while the other half [of humanity] ran' (Gilman 1966:330). It was sexual inequality not sexual equality which in her view posed the real threat to further evolutionary advance.

Blackwell's and Gilman's arguments rested on the assumption that men's and women's psychological faculties should equal each other in value. Spencer and Darwin, as we have seen, believed that as things currently stood these faculties were clearly unequal. Other evolutionary theorists, however, asserted that femininity was equal in worth with masculinity. 'To dispute whether males or females are the higher,' wrote Geddes and Thomson, 'is like disputing the relative superiority of animals or plants. Each is higher in its own way, and the two are complementary' (Geddes and Thomson 1890:249). The biological argument provided by these two writers for the essential equivalence of the two sexes – as 'equals but not identicals' (to use a phrase of Blackwell's, 1875 : 11) – was to prove peculiarly attractive to those feminists who sought equality on the basis of a celebration of women's supposedly specific 'feminine' traits. And this was despite the fact that their arguments had also been repeatedly used to advance the cause of the anti-feminists.

The extremely prominent settlement worker and suffragette, Jane Addams, was particularly influential in spreading Geddes's ideas in America (Conway 1970). Like Geddes, she argued that women had a qualitatively different character from men. Women should be given the vote because government stood to gain enormously from their particular, feminine skills:

> 'The very multifariousness and complexity of a city government demands the help of minds accustomed to detail and variety of work, to a sense of obligation for the health and welfare of young children, and to a responsibility for the cleanliness and comfort of others.' (Addams 1907 : 606)

Geddes and Thomson had claimed that women's functions were anabolic and were concerned with feeding and conserving the race. Addams (1922), in her turn, claimed that these feminine qualities could and should be used in the furtherance of social progress. Women, she said, as a result of their time-immemorial concerns with feeding others, could speed the cause of peace by using these talents to develop international relief efforts.

Many (e.g. Ellis 1929; Kraditor 1965) have attributed the success of the suffrage movement to the triumph of this wing of the women's movement, to the fact that suffragettes like Addams pleaded their cause, not in terms of women's common humanity with men, but in terms of their supposed essential dissimilarity, in terms of their particular feminine talents.

Certainly some of those who were most instrumental in securing the vote for women later claimed that their actions in this respect had been inspired by the belief that women had special feminine skills to contribute to government. An Australian cabinet minister, for instance, claimed that the biological case made by Geddes and Thomson for the equality, though not identity, of the sexes had influenced his introduction of a Bill giving women the vote in his country – a Bill which was the forerunner of similar legislation in England and America (Boardman 1978).

Such essentialist arguments, as I have called them (i.e. arguments claiming that women have been endowed with essentially different character traits from men) do not, however, serve the women's cause in the long term. Indeed, the prominence of these arguments during the women's suffrage campaign may well have contributed to the demise of feminism in the years following the achievement of the suffrage. Once they had secured the vote many of these women were content to confine their supposedly biologically given feminine skills to those spheres that society proclaimed to be particularly suited to them: to the home and to occupations like social work, teaching, and nursing. Despite the feminist consciousness produced in some middle-class women by their new-found college experience, many were to go on into careers which fostered the notion that women should occupy a different – feminine – sphere from men (Conway 1974). Ellen Richards, for instance, used her chemistry training to found the discipline of Home Economics – a discipline that was to reinforce much of the twentieth-century feminine mystique of housework (Ehrenreich and English 1979).

Social Darwinism criticized

In confronting the biological arguments of Maudsley and Clarke (see Chapter 2 above) some feminists had conceded that biology might, as these anti-feminists proclaimed, have rendered women essentially different from men as regards intellectual work (e.g. Higginson 1874; Jackson 1874). Others pointed out that these arguments were patently absurd: 'If,' said one writer, 'boys and girls may feed their bodies with the same things, from the same table, by what logic can we separate the same boys and girls in the process of mind-feeding?' (Stevenson 1881: 151).

Similarly the social Darwinist arguments adopted by feminists like Blackwell, Gilman, and Addams are equally to be rejected, for they were flawed for exactly the same reasons as the arguments of the social Darwinist anti-feminists. Spencer's arguments about the biological determinants of the division of labour between the sexes were as I have indicated (see pp. 34–5 above), circular and in no way justified his claims that existing human sex roles had been determined by evolution. Nor, therefore, did they justify either

Blackwell's or Gilman's claims about the biological determinants of the traditional division of labour between the sexes.

Geddes's and Thomson's premises about the biological differences between the sexes also did not warrant their conclusion that biology had dictated existing psychological and social inequalities between the sexes. Although they claimed to have deduced the different mental characters of men and women from the assumption of sex differences in cell metabolism these 'deductions' were, in fact, entirely spurious. They were dependent on these authors' prior conceptions about the character of masculinity and femininity. They were in no way independently derived from biology. As Helen Thompson, a research psychologist, pointed out at the time:

> 'Women are said [by Geddes and Thomson] to represent concentration, patience, and stability in emotional life. One might logically conclude that prolonged concentration of attention and unbiased generalization would be their intellectual characteristics. But these are the very characteristics assigned to men . . . Men, whose activity is essentially intermittent, and whose emotions are greater in variety and more unstable, are characterised by prolonged strains of attention and unbiased judgment. It may be true, but the proof for it does not appeal to one as very cogent. In fact, after reading the several expositions of this theory, one is left with a strong impression that, if the authors' views as to the mental differences of sex had been different, they might as easily have derived a very different set of characteristics.' (Thompson 1903:173–74)

Another writer of this period also pointed out that Geddes's and Thomson's theory rested on false analogy and that the biological character of the sex cells could have been just as well used by them to demonstrate 'the versatility, the fickleness and the weakness of men' and the 'seriousness and weight' of women (Finot 1913:135). Sexism, it seemed, not biology, had dictated Geddes's and Thomson's conclusions in this matter.

Nor was there any truth in Geddes's and Thomson's or Addams's claims that women had certain eternal, biologically given, feminine qualities. There is no clear evidence that girls and women today consistently demonstrate the traits stereotypically associated with femininity (Maccoby and Jacklin 1974). Even at the time at which

Addams wrote proclaiming women's eternal feminine virtues it was clear to many of her contemporaries that women were changing; and that these changes were being wrought in their psychology by the changes that were then occurring in their social situation. Helen Thompson, Addams's fellow Chicagoan, asserted for instance that 'There are, as everyone must recognize, signs of a radical change in the social ideals of sex' (Thompson 1903: 182). Masculinity and femininity, far from being the eternal and fixed characters claimed by Geddes and Thomson, were clearly not immune from the effects of historical change. As another writer of this time pointed out, women's 'essential and eternal type entirely escapes us . . . Times have changed, and women also. It is only the psychologists who have not altered' (Finot 1913: 192).

It has recently been suggested (Rosenberg 1975) that the different perspectives on the one hand of Thompson, and on the other of Blackwell, Gilman, and Addams on the woman question simply reflect differences in their social situation. Thompson's rejection of biological essentialism, it is said, reflected the fact that as a graduate student she had enjoyed equal status with her male colleagues and had therefore felt no need to define a separate feminine sphere for women. Blackwell and Gilman as mothers, and Addams as an organizer of a primarily female social work community, may, on the other hand, have felt the need to define a separate feminine sphere for women in the home and in the wider social community. That their different positions on the issue of women's role might have reflected differences in their social situation does not, however, vouch for the validity of these positions. Given that women's character has not remained eternally the same but has changed in the course of history, Thompson's rejection of social Darwinist accounts of the biologically given and, by implication, relatively eternal nature of femininity seems the more valid basis for analyzing the proper role of women in society.

Lastly, the social changes which had led to changes in women's destiny were not simply the automatic effect of a biologically given evolutionary process, as both feminist and anti-feminist social Darwinists suggested. They were the effect of historical development and of human agency. Improvements in women's situation occurred, in part at least, as the result of the struggles that men and women had conducted together on behalf of women's rights. The joint struggle of both sexes around this issue had been possible because, despite the claims of the social Darwinists, men and women had not been confined by biology to separate spheres of influence.

Evolutionist theory had often wrongly advocated, either explicitly or implicitly, a passive acceptance of any and all social developments that might affect women's role, on the grounds that since these developments were the product of evolution, they must therefore be progressive. This passivity was nicely expressed by one writer at the time who, in defending changes then occurring in women's status, maintained:

> 'Daughters are not revolting but *being evolved*, and the evolution may be looked upon with great placidity . . . . Certainly the restrictions which produce the feeble earth-bound Dodo, *must give way* to the freedom which will give the angels in our homes room to grow their six strong wings – two for personal dignity and beauty, two for spiritual elevation, and two with which to fly on serviceable errands for humanity' (Amos 1894: 515, 520. My emphasis.)

Women have never been simply given their freedom, they have repeatedly had to fight for it. The victories that women have achieved through the struggle for sexual equality demonstrates that women's destiny, as well as being a function of human history, is also a function of human agency; and that evolutionism, though valid as biological theory, is bankrupt as an account of the development of human society.

Notes

1 He had, he said, only included the sentence 'in order that no honourable man should accuse me of concealing my views'. Lack of sufficient evidence meant that it would have been 'useless and injurious to the success of the book' to have expanded the point any further at this stage (Darwin 1888, Vol. I: 75–6).

2 Darwin defined sexual, as opposed to natural, selection as follows:

> 'Sexual selection depends on the success of certain individuals over others of the same sex, in relation to the propagation of the species; whilst natural selection depends on the success of both sexes, at all ages, in relation to the general conditions of life. The sexual struggle is of two kinds; in the one it is between the individuals of the same sex, generally the males, in order to drive away or kill their rivals, the females remaining passive; whilst in the other, the struggle is likewise between the individuals of the same sex, in order to excite or charm those of the opposite sex, generally the females, which no longer remain passive, but select the more agreeable partners.' (Darwin 1896: 614)

3 He has, for instance, recently been awarded the prestigious Pulitzer Prize.
4 Wilson's theory has been cited, for example, by Richard Verral (1979) as support for the extreme right-wing doctrines of the National Front.
5 Spencer was not alone in becoming less generous on the matter of women's rights as these rights came to be more vigorously pursued in the 1860s. As feminist ideas came to take effect so other writers – e.g. Edward Clarke (see Walsh 1977) and James McGrigor Allan (1868) – also came to revise their previous arguments in favour of sexual equality.
6 The explanation of the causes of variation in species still poses problems for evolutionary theory. One interesting suggestion (Ho and Saunders 1981) is that, as well as being empirical, these problems are also, in part, epistemological and stem from the fact that evolutionary theorists generally adopt a mechanical, rather than a dialectical materialist framework of analysis.
7 They also claimed to deduce these psychological sex differences from the widely held view (e.g. Brooks 1879) that the female sex cell is the conservative, the male sex cell the progressive and innovative element in evolution.
8 Darwin recommended, for instance, that 'In order that woman should reach the same standard as man, she ought, when nearly adult, to be trained to energy and perseverance and to have her reason and imagination exercised to the highest point' (Darwin 1896: 565).
9 J.D. was, in all probability, John Dewey (Ehrlich 1975).

Four

Sociobiology on the relations between the sexes

In the previous chapter I looked at some of the historical precursors of sociobiology. I shall devote the next two chapters to a consideration of the arguments of sociobiology itself, at least as they relate to the position of women in society. In this chapter I shall examine the sociobiological claim that certain features of human sexual relations – namely, women's traditional role in childcare, the sexual double standard, and the battle of the sexes – are given by biology. I shall then turn, in the next chapter, to a consideration of the thesis that men's dominance of political and social life is determined by innate aggression.

The 'selfish gene' and the 'battle of the sexes'

One of the basic premises of sociobiology is that individuals act so as to maximize their 'inclusive fitness'.[1] That is, sociobiologists adopt the thesis put forward by biologist W. D. Hamilton (1964) that individuals act so as to maximize the chances of their genes surviving by promoting their own welfare and that of their relatives who share their genes. Behaviour, in their view, is essentially governed by genetic self-interest.

The fact that individuals – both humans and other animals – also act altruistically and, therefore, apparently unselfishly poses a problem for any such theory of behaviour. In keeping with Hobbes's timeworn method of attempting to resolve this dilemma, Hamilton and his followers have sought to reduce altruism to self-interest. Hamilton proposes that individuals act altruistically in so far as they can promote the well-being of their genes via helping their relations. Altruism, in Hamilton's account of it, thus turns out to be simply another case of genetic self-interest. The care that parents bestow on their children is, in his view, a special case of this kind of altruism. Parents, he says, protect their offspring even to the extent of putting their own lives at risk, because they thereby increase the chances of their genes surviving via their offspring.

Subsequently, Trivers has pointed out that even though parental care ultimately serves one's genetic self-interest, it also places a limit on the number of offspring one can rear to reproductive maturity. Like Hamilton, and indeed like Malthus, Trivers assumes that individuals act so as to maximize their reproductive success. Whereas Malthus came to regard this process as limited – at least among humans – by 'moral restraint, vice and misery' (Malthus 1960 : 477), Trivers regards it as limited by 'parental investment' which he defines as 'any investment by the parent in an individual offspring that increases the offspring's chance of surviving (and hence reproductive success) at the cost of the parent's ability to invest in other offspring' (Trivers 1972 : 139).

Evolution, he says, has resulted in females 'investing' more food resources for their offspring in the sex cell than do males. The egg is much richer in food reserves than the sperm. This biologically given imbalance between the sexes in parental investment at conception results, he says, in a similar imbalance between them once their offspring are born. Since, he argues, females invest more in their offspring at conception, they will also continue to invest more in them postnatally in terms of childcare. In this context he states that the female's 'initial very great investment commits her to additional investment more than the male's initial slight investment commits him' (Trivers 1972 : 144).

Although much of Trivers's (1972) article is devoted to the discussion of sex differences in 'parental investment' among birds, he also maintains that this same argument applies to human sex roles in childcare. He suggests that biologically given differences in the food reserves of the male and female sex cells dictate the traditionally unequal division of childcare between the sexes, and that this also explains why it is that men are more sexually promiscuous than women:

> 'After a nine-month pregnancy, a female is more or less free to terminate her investment at any moment but doing so wastes her investment up until then. Given the initial imbalance in investment the male may maximize his chances of leaving surviving offspring by copulating with and abandoning many females, some of whom, alone or with the aid of others, will raise his offspring.' (Trivers 1972 : 145)

He further maintains that, since biology has ordained that both males and females act so as to maximize their own individual reproductive success, relations between them will be essentially those of mutual exploitation, 'that even when ostensibly cooperating in a joint task male and female interests are rarely identical' (Trivers 1972 : 174). This point is put even more clearly by Richard Dawkins, one of the main popularizers of Trivers's theory. Since the behaviour of both sexes is determined by the ethic of 'the selfish gene', each sex will therefore try to get the other to 'invest' more parentally in their joint offspring thus leaving them free to produce and rear further offspring:

> 'If one parent can get away with investing less than his or her fair share of costly resources in each child . . . he will be better off, since he will have more to spend on other children by other sexual partners, and so propagate more of his genes. Each partner can therefore be thought of as trying to exploit the other, trying to force the other one to invest more.' (Dawkins 1976 : 151)

And, on this basis, he characterizes the relations between the sexes as a 'battle'.

Both Trivers and Dawkins go on to claim that biologically given sex differences in parental investment at conception, together with the shared concern of both sexes to maximize their reproductive success, have resulted in a number of biologically given strategies whereby each sex attempts to exploit and outwit the other. Since biology has dictated that females invest more in their offspring at conception and once they are born, it has, they argue, been adaptive for males to evolve a strategy of 'philandery'. Evolution, they say, has favoured those males who desert their mates for others once they have impregnated them since in this way males can yet further increase their reproductive success. Such philandery, say these socio-biologists, will not reduce the chances of their already sired offspring surviving to reproductive maturity for females can, on the basis of their prior parental investment, be relied upon to rear these offspring.

Faced with this male strategy of philandery, say Trivers and Dawkins, it has befitted females to evolve the counter-strategy of 'coyness'. By acting coyly, they say, females can give themselves an interval prior to mating in which to assess the likelihood that their prospective mate will provide childcare for their offspring once they are born. Males, says Dawkins, may then reply to this strategy by acting 'faithfully' to achieve mating, and only then act the 'philanderer', desert their mate, and seek another whereby to increase their reproductive success still further. Females, in turn, may retaliate against such desertion by trying to dupe another male into providing care for their offspring and thus reduce the amount of parental investment they have to provide. Finally, in order to pre-empt this strategy, the male, according to Trivers and Dawkins, may sequester his female partner away from other males prior to mating thereby ensuring that the offspring she produces are his and that he does not waste parental investment (in the form of childcare) on offspring that do not contain his genes.

Trivers suggests that these strategies, by which each sex tries to outwit the other in order to maximize its own inclusive fitness and minimize its parental investment, are the product of evolution. Dawkins, following this evolutionary logic, proposes that the immediate biological determinants of these strategies are genetic. Both authors thus imply that women's subordination – at least in so far as it is expressed in the unequal distribution of childcare between the sexes and in the sexual double standard – is biologically determined.

Trivers's theory concerning the biological determinants of human sex roles has been enormously influential. His theory has been popularized by Dawkins and by E. O. Wilson. It informs the teaching of biology in American high schools (Lowe 1978) and is promoted by many widely used undergraduate textbooks (e.g. Barash 1977; Daly and Wilson 1978). Furthermore, the economistic terminology of parental investment theory is even used by many avowedly feminist scholars (e.g. Tanner and Zihlman 1976; Rossi 1977). Despite its popularity, however, Trivers fails to make good his claims regarding the biological determinants of human sex roles. I shall outline some of the flaws in Trivers's argument in the next sections of this chapter.

Darwinian and sociobiological evolutionism

First, it should be pointed out that sociobiology, while trading on the respectability of Darwin's theory of evolution, also departs from it in several respects. In this it is like social Darwinism. Just as the differences between social Darwinism and Darwinian theory were glossed over by many social Darwinists when they appealed to the authority of Darwin to bolster their conservative viewpoint on the woman question (see Chapter 3 above), so differences between modern sociobiology and Darwin's theory of evolution are also glossed over in current sociobiological accounts of the evolutionary determinants of traditional sex roles.

Trivers and Dawkins rest their case for the biologically given exploitativeness of human sexual relations on the premise that individuals act so as to maximize the number and chances of their genes surviving to maturity. They argue that selection favours those individuals who *maximize* their reproductive success. Darwin, on the other hand, claimed that selection favours those individuals who succeed in producing, not the maximum possible number of off-spring, but simply more offspring *relative* to others of their species (Sahlins 1976; Quadagno 1979).

Furthermore, Darwin's theory of evolution implies that species characters are not fixed but change as the effect of chance variation and of the selection of those variations that prove relatively well adapted to prevailing environmental conditions. Trivers and Dawkins, by contrast, suggest that certain behavioural adaptations, such as male philandery, are fixed for all time, and that other behaviours have been evolved teleologically to meet this biological given. Philandery, they say, is the optimum strategy for males to pursue in order to maximize their reproductive success, and has, they argue, therefore been selected for in the course of evolution and is now biologically fixed. Female coyness and male attempts to secure female chastity have been evolved in order to meet this biological given.

It might be argued that it is legitimate to depart from Darwinian theory in the above ways. After all, Darwin's theory is over a hundred years old and Trivers's theory, it may be said, represents a legitimate advance on nineteenth-century evolutionary theory. But as Sahlins points out, even if one accepts Trivers's first departure from Darwinian evolutionary theory (i.e. his premise that individuals act to maximize their reproductive success), this does not warrant his conclusion that male philandery is biologically inevitable. 'There is,' says Sahlins,

> 'no showing on Trivers's part that the reproductive advantages of desertion for the male are any greater than fitness losses he is liable to incur in competition – not to mention that abandonment of his one-time consort reduces her chances of raising offspring. Without additional assumptions or observations, there is no basis at all for supposing that this kind of exploitation of females maximizes the individual male's reproduction, hence is "selected for".' (Sahlins 1976:90)

Since Trivers fails to produce any such 'additional assumptions or observations' he fails to make good his argument that male philandery is an optimal, and hence a fixed behavioural strategy under all conditions. Nor, therefore, does he succeed in justifying his further departure from Darwinian theory, namely his teleological argument that other sexual strategies have been evolved in order to serve the function of outwitting male philandery. Trivers thus fails to justify these details of his account of sex roles even in terms of his own evolutionary theory, let alone in terms of the Darwinian theory on which this account is purportedly based. I shall now go on to argue that his account of human sex roles, although phrased in apparently purely biological terms, in fact relies on certain social – not biological – presuppositions.

The culturally relative presuppositions of parental investment theory

The attempt to derive the character of human behaviour from the premise that it is ultimately governed by self-interest is, of course, not new to sociobiology. As I have already pointed out, Thomas Hobbes advanced very much the same kind of theory. Just as Hamilton (1964) and Trivers (1971) now claim that altruism is motivated by genetic self-interest, so Hobbes claimed in 1651 that unselfishness is ultimately reducible to selfishness: 'For,' he said, 'no man giveth, but with intention of Good to himselfe' (Hobbes 1950:125).

It may well be argued that even though Trivers does not acknowledge the roots within Hobbesian social theory of his avowedly biological theory of animal behaviour, the fact that such roots exist does not invalidate his claim that animal behaviour is indeed governed by self-interest. After all there is good precedent in Darwin's own theory for using social theory to derive biological theory. Darwin quite explicitly and openly acknowledged that his account of natural selection had been influenced by Malthus's theory of population. He wrote as much in his *Autobiography*:

> 'In October 1838, that is, fifteen months after I had begun my systematic inquiry, I happened to read for amusement "Malthus on Population", and being well prepared to appreciate the struggle for existence which everywhere goes on from long-continued observation of the habits of animals and plants, it at once struck me that under these circumstances favourable variations would tend to be preserved, and unfavourable ones destroyed. The result of this would be the formation of a new species. Here then I had at last got a theory by which to work.' (Darwin 1888, Vol. I: 67–8)

The fact that Darwin had used Malthus's social theory in order to arrive at his account of the biological processes of evolution did not invalidate his theory, qua biological theory. As Engels pointed out in criticism of Dühring's rejection of Darwinism on the grounds of its apparent equivalence with Malthusianism: 'no Malthusian spectacles are required to perceive the struggle for existence in nature . . . the struggle for existence can take place in nature, even without any Malthusian interpretation' (Engels 1976: 86).[2]

Both Darwin and Engels did, however, object to the circular use of this account of nature by social Darwinists in order to justify the social attitudes from which this biological theory had been derived in the first place. Darwin was amused, but dismissive, of such foolish uses of his theory. 'I have,' he wrote to Lyell, 'received in a Manchester newspaper rather a good squib, showing that I have proved "might be right", and therefore that Napoleon is right, and every cheating tradesman is also right' (Hofstadter 1955: 85). Engels denounced the same type of reasoning as obvious 'childishness'. Social Darwinism, he pointed out, involved the following circularity:

> 'The whole Darwinian theory of the struggle for existence is simply the transference from society to animate nature of Hobbes's theory of the war of every man against every man and the bourgeois economic theory of competition, along with the Malthusian theory of population. This feat having been accomplished . . . the same

> theories are next transferred back again from organic nature to history and their validity as eternal laws of human society declared to have been proved.' (Engels 1875 : 198)

Sociobiology is open to similar attack. Like social Darwinism, it also relies on circular reasoning. It uses terms derived from present human society to characterize animal behaviour and then uses this characterization to justify, in biological terms, the human society from which that characterization was derived in the first place. Trivers and Dawkins describe sexual behaviour among animals in terms of 'philandery', 'coyness', and 'dishonesty'; that is, in terms derived from our own particular human society and from the sexual double standard that obtains in it. They then use this description of animal behaviour in order to attempt to legitimate in biological terms the social order and its sexual double standard, from which this description was itself derived.

It is a measure of the extent to which sociobiologists are unaware of the ways in which they rely on culture-bound assumptions about human sexual relations to characterize animal behaviour that they often regard these assumptions as constituting independent evidence for the validity of their biological explanation of human sex roles. Daly and Wilson, for instance, argue that

> 'Evolutionary biology predicts differences in the behavioural inclinations of the sexes, and human behaviour fits the predictions. Women are indeed more selective in choosing mates than men. Men indeed generally react more strongly to spousal infidelity. Women are generally courted by men. Men generally appear to be more polygamously inclined than women.' (Daly and Wilson 1979 : 17)

Although Darwin used social theory (i.e. Malthus's theory of population) to arrive at his account of biological evolution, he also provided a wealth of independent biological observations to corroborate this account. Sociobiologists, on the other hand, as we have seen, often use the same social attitudes to corroborate aspects of their theory as they use to derive that theory in the first place. This is not to deny that writers like E. O. Wilson provide a wealth of detailed observations on animals in order to illustrate their claims about animal behaviour. They have, however, singularly failed to provide any systematic or comprehensive review of the ethnographic and historical data on human societies in order to make good their claim that certain features of these societies are also dictated for all time by biology. Examination of this data indicates that sociobiology, at least its account of the sexual double standard among humans and of the exploitative character of human sexual relations, is not, as it claims, universally and eternally valid.

Consider, first, the sociobiological account of the sexual double standard. Trivers and his followers point out that the biological fact that women bear children means that they have more certainty than men as to who are their biological offspring. Sociobiologist David Barash stresses, for instance, that

> 'The one commonality shared by Alaskan Eskimos, Australian aborigines, African Bushmen and Wall Street businessmen is their biological heritage; one aspect of that heritage is that males of virtually all animal species must have less confidence in their paternity than females have in their maternity.' (Barash 1977:300)

Sociobiologists argue that this biologically given sex difference in certainty regarding offspring dictates the sexual double standard. Men, they say, insist on female sexual chastity in order to ensure that the offspring they produce are genetically theirs, and that they do not waste parental investment on offspring that do not contain their genes.

Certainly men's concern lest they invest parentally in offspring which are not their own, at least in the sense of bequeathing property to them, has accounted for the insistence among the propertied classes on female chastity (see Chapter 10 below for further discussion of this point). However, it stands to reason that concern about this kind of parental investment cannot be the direct cause of the sexual double standard among the non-propertied classes, still less among societies that have no form of heritable property. Moreover, there is little evidence that anxiety about any other kind of parental investment (such as childcare) causes an insistence on female chastity. There are, therefore, no grounds for the sociobiological thesis that such anxiety, arising from the biologically given fact that men are less certain than women of their offspring, universally determines the sexual double standard in which women, but not men, are required to be sexually chaste.

Consider next the sociobiological claim that biology dictates that the relations between the sexes be essentially those of mutual exploitation, in which each sex attempts to secure the childcare services of its mate so as to enable it to promote its own genetic self-interest. Dawkins, as we have seen, argues that males feign 'faithfulness' to secure the services of females in the form of the bearing and rearing of their offspring (i.e. of their genes) and only then act the 'philanderer' so as to sire yet more offspring by other females. Females, he says, counter this strategy either by acting 'coyly' or by duping males into looking after offspring which are not their own. Both strategies, he claims, are now genetically determined because they serve to secure the childcare services of males so that the female can bear and rear yet more offspring and thus serve her own genetic self-interest.

Although sociobiologists argue that these strategies are given by biology and thus imply that they are eternally fixed features of human sexual relations, their argument in fact relies on an assumption which is not eternally valid, but instead only holds true of certain sections of some societies, and then only of particular stages in the historical development of those societies. In arguing that genetic self-interest leads individuals to try to secure the childcare services of the opposite sex, sociobiologists presuppose that labour power (in the form of childcare service) is freely alienable and that it can be appropriated on a free market basis by other individuals (particularly by members of the opposite sex). But although this presupposition might generally hold true in our society it does not generally hold good in other societies. Labour power is not freely alienable, for instance, in feudal or slave-based societies in which much of the labour force is 'tied to the land, or to the performance of allotted functions, or (in the case of slaves) to masters' (Macpherson 1962 : 49). The sociobiological claim that biology has dictated that sexual relations are eternally and universally marked by mutual exploitation – by a 'battle of the sexes' – thus turns out to presuppose certain forms of relatedness between individuals, forms that are by no means universal.

This flaw in such accounts of human nature has been pointed out by the political theorist Crawford Macpherson in his discussion of Hobbes's theory about the exploitativeness of human social relations. Like the sociobiologists of today, Hobbes claimed that the premise that individuals act out of self-interest implies that humans will naturally attempt to secure the services of others in order to help them promote their own selfish interests. People, he said, naturally have 'a perpetuall and restlesse desire of Power after power, that ceaseth onely in Death' (Hobbes 1950: 79), a 'restlesse desire' to secure the labour power – the 'assistance, and service' (Hobbes 1950: 70) – of others. As Macpherson shows, however, Hobbes's argument in this respect presupposes that individuals are free to grant their services to others, that, as Hobbes himself puts it, a 'man's Labour also, is a commodity exchangeable for benefit' (1950: 209). Unlike today's sociobiologists, Hobbes was perfectly candid in making this presupposition. He asserted that 'The Value, or Worth of a man, is as of all other things, his Price; that is to say, so much as would be given for the use of his Power' (Hobbes 1950: 70). He did not, however, point out that this presupposition only held true of certain sections of his own society. At the time at which he was writing, less than half the men were full-time wage earners (Macpherson 1962). The majority were not therefore disposing of their labour on the kind of free market basis presupposed by Hobbes in his account of human behaviour. In sum, the description of the 'Naturall Condition of Mankind' (Hobbes 1950: 101), provided by Hobbes and by present-day sociobiology, is not given by nature, but is instead dependent for its validity on specific historical contingencies – contingencies that did not fully obtain in Hobbes's time and that barely obtained at all in the feudal society that preceded it.

Human sexual relations: exploitative or cooperative?

I shall conclude this chapter by comparing the sociobiological characterization of the relations between the sexes with an earlier social Darwinist characterization of them. The claim of sociobiology that the relations between the sexes are essentially exploitative is a far cry from the social Darwinist claim of Geddes and Thomson (see Chapter 3 above) that biology has ordained that these relations be essentially cooperative in character. The difference between these two accounts of human sex roles parallel their authors' different positions regarding Darwinian theory and its claims about the centrality of 'the struggle for existence' as a motor of evolutionary change.

Geddes and Thomson were unhappy with the Darwinian idea that evolution necessarily proceeds through struggle. Like those writers against whom Malthus had originally written his *Essay on the*

Principle of Population, Geddes and Thomson believed in the 'perfectibility of man' and in the inevitability of social progress. Whereas Darwin explicitly rejected the claim that evolutionary progress was inevitable, Geddes and Thomson adhered to this claim and, on this ground, rejected Darwin's view that evolutionary development resulted primarily from the struggle between individuals for existence. On this matter they asserted:

> 'Each of the greater steps of progress is in fact associated with an increased measure of subordination of individual competition to reproductive or social ends, and of inter-specific competition to cooperative association . . . The ideal of evolution is indeed an Eden; and although competition can never be wholly eliminated, and progress must thus approach without ever completely reaching its ideal, it is much for our pure natural history to recognise that "creation's final law" is not struggle but love.' (Geddes and Thomson 1890 : 286)

Whereas Geddes and Thomson explicitly took issue with Darwin's theory of sexual selection and the prominence he gave to competitive struggle, Trivers's (1972) account of sex roles in terms of parental investment, competitiveness, and genetic self-interest is equally explicitly part of a movement to revive enthusiasm for Darwin's theory of sexual selection.

The contrast between their different views of nature has its forerunner in a similar contrast which marked the biological viewpoints that preceded and those that followed the publication of *The Origin of Species* and *The Descent of Man*. Engels described this contrast as follows: 'Before Darwin, the very people (Vogt, Buchner, Moleschott, etc.) who now see nothing but the *struggle* for existence everywhere were stressing precisely the *cooperation* in organic nature' (Engels 1875 : 197. Engels's emphasis). Nevertheless as Engels rightly went on to point out: 'Both conceptions have a certain justification within certain limits, but each is as one-sided and narrow as the other. The interaction of natural bodies – whether animate or inanimate – includes alike harmony and collision, struggle and cooperation' (Engels 1875 : 197). Similarly, Geddes's and Thomson's and Trivers's descriptions of nature and of the relations between the sexes are each partially correct; both nature in general, and the relations between the sexes in particular, are marked by cooperation and by struggle. But in attempting to reduce one aspect of these relations – either that of cooperation or of conflict – to the other contradictory aspect, these biologists fail to do justice to both aspects of these relations.

It might be argued, at this point, that contrary to what I argued in the previous chapter Geddes's and Thomson's theory hardly constitutes an essentially similar account of human sex roles to that provided by sociobiology. In fact, however, Trivers's theory is similar to that of Geddes and Thomson. Like these earlier social evolutionists, Trivers also attempts to deduce the character of human sex roles from the biology of the sex cells, a deduction which, as in their case, turns out to be unwarranted.

Geddes's and Thomson's claim that masculinity and femininity, and therefore the existing division of labour between the sexes, is given by sex differences in the metabolism of the sex cells rested, as we have seen (Chapter 3 above), on false analogies. Trivers's claim that the existing division of childcare between the sexes is the effect of differences in the food reserves of the female and male sex cells is also unjustified. He assumes that the greater initial investment of food reserves by the female in her eggs dictates her greater parental investment in offspring once they are born. Dawkins, however, who entirely subscribes to this assumption of Trivers as applied to females, describes it as 'fallacious economics' when applied to males:

> 'A business man should never say "I have already invested so much in the Concorde airliner (for instance) that I cannot afford to scrap it now." He should always ask instead whether it would pay him in the future, to cut his losses, and abandon the project now, even though he has already invested heavily in it. Similarly, it is no use a female forcing a male to invest heavily in her in the hope that this, on its own, will deter the male from subsequently deserting.' (Dawkins 1976 : 162)

But if prior parental investment does not ensure future parental investment where males are concerned, why should it ensure future parental investment where females are concerned? Even if we accept the terms of parental investment theory, and agree to characterize the female as 'investing' more in her eggs than do males in their sperm, this is no reason, in Dawkins's and Trivers's economistic terms, why she should continue to invest in her offspring once they are born. But this is precisely what these biologists have to assume when they try to justify the traditional division of childcare between the sexes in terms of sex differences in parental investment in the sex cells at conception.

In sum, these biologists have failed to demonstrate that traditional sex roles in childcare are dictated by biologically given differences in the sex cells. And, as I argued earlier in this chapter, they have also failed to demonstrate that nature has dictated the sexual double standard which currently prevails within our society. Nor have they proved, as they claim to have done, that the relations between the sexes, as they define them, have been given for all time by biology. Their explanations of these features of human sexual relations presuppose certain forms of human social relationship that have not always existed. The validity of these explanations is not given, as sociobiologists imply, by unchangeable and eternal features of human biology. It is instead dependent on the existence of particular forms of human society – of societies in which surplus wealth is conferred on individual men to invest in their children, and in which labour power is available on a free market basis – forms of society which, although they have existed and continue to exist in many parts of the world, have not always prevailed and need not always prevail in the future.

Notes

1 There is some disagreement between sociobiologists over the definition of inclusive fitness. Some, like David Barash, define it as 'The sum of an individual's fitness as measured by personal reproductive success and that of relatives, with those relatives devalued in proportion to their genetic distance, i.e., as they share fewer genes' (Barash 1977: 329). E. O. Wilson, by contrast, defines it as applying only to indirect relatives, not to 'direct descendants' (Wilson 1975a: 586).

2 A similar debate continues between writers on the left today: between those (e.g. Young 1977) who criticize capitalist science, including Darwinian evolutionism, in the belief that it is nothing but ideology, and those (e.g. Rose 1979) who argue that although science is indeed influenced by ideology it also has an independent 'field' (Rose 1979: 286).

Five

Physical strength, aggression, and male dominance

Having examined the sociobiological thesis that certain features of the personal relations between the sexes are dictated by biology, I shall now go on to examine the claim that men's dominance of social and political life is determined by innate aggressiveness. This is a claim which, as we shall see, sociobiologist E. O. Wilson advances, and it has also been put forward by a number of other recent writers on sex roles, writers who do not explicitly label themselves sociobiologists.

The argument that male dominance is the effect of biologically given sex differences in aggression, like Trivers's claims about the biological determinants of human sex roles, rests on a historically specific social theory. It rests on the doctrine that individual success is the result of competition. And just as the social doctrines on which Trivers rests his thesis have their roots in Hobbes, so too the argument that male dominance is the effect of male aggression is based on a social doctrine promulgated by Hobbes: Hobbes claimed that competitiveness was given by the 'naturall condition of mankind', that it was the result of the natural inclination of all individuals to pursue their own self-interest, and that therefore 'in the nature of man, we find three principall causes of quarrell. First Competition; Secondly, Diffidence; Thirdly, Glory' (Hobbes 1950:101, 103).

Before considering the recent version of this doctrine, namely that competitiveness is rooted in the sex hormones, I shall first consider some nineteenth-century versions of this doctrine as it relates to the subordinate status of women in society. That is, I shall look first at the argument that male dominance is the result of biologically given sex differences in physical strength and the capacity for violence.

On strength, violence, and male dominance

Hobbes had the following to say of the relation between men's capacity for violence and the subjugation of women: competitiveness between men resulted in them using 'violence, to make themselves Masters of other men's persons, wives, children, and cattell' (Hobbes 1950:103). Like Hobbes, Spencer also claimed that might makes right, that male strength and hence men's greater capacity for violence had given them the right to dominate women. Although he anticipated that in 'domestic life, the relative position of women will doubtless rise', he deemed it unlikely that women would ever attain 'absolute equality' with men for, he argued, 'in the moral relations of married life, the preponderance of power, resulting from greater massiveness of nature, must, however unobtrusive it may become, continue with the man' (Spencer 1898:768).

Henry Maudsley, though no advocate of women's rights, pointed out the speciousness of this kind of reasoning. Men's superior strength was not, he said, a cause of women's subordination 'any more than superiority of muscular strength has availed to give the lion or the elephant possession of the earth' (Maudsley 1874:479). Nevertheless writers continued to argue, as Spencer had, that men's physical strength explained their dominance of women. Like Spencer, they attempted to justify their belief in the inevitability and justice of patriarchy in biological terms.

The development of European colonialism and of American Westward Expansion brought anthropologists into contact with apparently non-patriarchal societies. Such contacts, in conjunction with the fact that traditional sex roles were being questioned at home, threatened to undermine the long held belief that human societies were naturally and eternally patriarchal in structure (Fee 1974). Some writers responded to this threat to the ideology of patriarchy by asserting that although women appeared to have more power in some 'primitive' societies than they did in 'civilized' societies, the underlying structure of these societies was also essentially patriarchal. Patriarchy, they said, had been given for all time by the fact that men were stronger than women. The American sociologist W. I. Thomas, for example, asserted that although mother-right (i.e. matrilineal descent and matrilocal residence) obtained in some societies, these societies were essentially patriarchal, just like those in which father-right obtained. Patriarchy, he said, was universal, and he claimed it was the inevitable effect of the superior strength of men:

> 'In view of his superior power of making movements and applying force, the male must inevitably assume control of the life-direction of the group . . . there has never been a moment in the history of society when the law of might, tempered by sexual affinity, did not prevail. We must, then, in fact, recognize a sharp distinction between the law of descent and the fact of authority. The male was everywhere present in primitive society, and everwhere made his force felt.' (Thomas 1898 : 761–62)

Others were not so convinced by this biological determinist argument. The status of women was clearly not universally the same. Women's status was obviously different in 'primitive' societies. Furthermore women's status in 'civilized' society was also clearly changing. Such cross-cultural and historical variations in sex roles cast considerable doubt on the validity of attempts to explain these roles as eternal, as given by fixed biological sex differences in force or strength.

It had already been pointed out that 'The *a priori* character of the family organisation as derived from the unequal distribution of strength between the man and the woman, are far from being realised in actual life' (Jacobi 1877 : 21). As industrialization advanced it became increasingly clear that sex differences in physical strength were no longer the cause of inequality between the sexes. Although the physical strength of men might have been responsible for their domination of women in the past, it hardly seemed to constitute an adequate explanation of their continuing domination of women in industrial society. Indeed, some writers gloomily warned that by rendering sex differences in physical strength irrelevant in productive work, industrialization would bring about the collapse of patriarchy.

The American sociologist Thorstein Veblen was typical in this respect. Patriarchy, he speculated, had developed in the first place out of inter-group warfare in which high value was placed on men's greater, biologically given fighting capacity. He suggested that this has resulted in an 'attitude of mastery on the part of the able-bodied men in all their relations with the weaker members of the group, and especially in their relations with women' (Veblen 1899: 506–07). Although patriarchy had in his view subsequently become sanctioned by tradition, it was now in the process of being overthrown by industrialization: its decline was, he said, particularly marked 'among the classes immediately engaged in modern industries' (Veblen 1899: 510). He went on to bemoan the effects of industrial development on patriarchal authority: 'The days of its [patriarchy's] best development are in the past, and the discipline of modern life – if not supplemented by a prudent inculcation of conservative ideals – will scarcely afford the psychological basis for its rehabilitation' (Veblen 1899: 511).

Others welcomed rather than lamented these effects of industrialization. Engels, as we shall see (Chapter 10), was hopeful that industrialization, in permitting the equal participation of women with men in social production, would pave the way for their emancipation. Nor was this view confined within marxism. Many of Engels's non-marxist contemporaries likewise welcomed the potentially liberating effects of industrialization on women's lives. Hitherto, said one such writer, 'the keenness of the struggle for life . . . and resulting therefrom the pre-eminence of physical force' had made women's lesser physical strength an 'insuperable obstacle' to their progress (Bulley 1890: 1). But, she argued, the way was now open for women's advance, since industrialization had rendered social power no longer dependent on physical force.

Today, many feminists retain the optimism of these nineteenth-century writers as to the potentially liberating effects for women of technological development. Dorothy Dinnerstein, for instance, claims that 'As current technology advances and spreads, size and muscle . . . have less and less to do with power [of men] to bully [women]' (Dinnerstein 1976: 39). She believes that now that technology has removed these physical obstacles to women's emancipation the only real barrier that remains is psychological. Others (e.g. Mitchell 1973) are, however, more pessimistic. They claim that nineteenth-century writers were wrong to predict that technological progress would help towards the emancipation of women. Although they admit that this advance did indeed mean that women could work alongside men in industry, they argue that this simply resulted in women becoming equal victims with men of capitalist industry and only barely advanced their emancipation.

The fact that, despite technological advance, men still do use their superior physical strength to subdue women through rape and wife abuse has once more served to focus attention on the role of male violence in the development and perpetuation of patriarchy. Susan Brownmiller, discussing rape, goes so far as to suggest that:

> 'It seems eminently sensible to hypothesise that man's violent capture and rape of the female led first to the establishment of a rudimentary mate-protectorate and then sometime later to the full-blown male solidification of power, the patriarchy.' (Brownmiller 1975 : 17)

And, writing about wife battering, other writers similarly claim that female subordination was initially a consequence of their lesser physical strength. They argue that in the beginning of human history 'Strong males naturally dominated the weaker females. Man literally took his wife and ruled her by physical force . . . Violence, or the threat of violence, kept women in their place' (Langley and Levy 1977 : 44–5). In this these writers reiterate the claim of some of the earlier anthropologists (e.g. McLennan 1865; Commons 1899) who located wife-capture as one of the origins of patriarchy – a claim which, I shall argue (see Chapter 10 below), is of dubious validity.

Unlike these earlier anthropologists, however, many of today's writers on rape and wife abuse regard the relation between men's greater physical strength and their social authority as being socially and historically contingent rather than biologically inevitable. In this they are more like the early anthropologist Malinowski than like, say, W. I. Thomas or John McLennan before him. Although Malinowski argued that woman's oppression, at least her 'drudgery' in Aborigineal society, was the effect of the 'compulsion of the weaker sex by the "brutal half of society" ' (Malinowski 1963 : 288), he also went on to point out that this was a sociological not a physiological matter – that the physical constraint of women by men was not given by biology but depended on social sanction. Similarly, those concerned with the problem of intersexual violence in our own society point out that its occurrence is a function, at least in part, of social and legal practice and is not simply determined by biology (e.g. Edholm, Harris, and Young 1977; Tomes 1978).

Faced with the ways in which industrial progress threatened to undermine nineteenth-century justifications of male authority, many writers argued then that even if men's greater physical strength did not *automatically* result in their having greater social authority than women, it *should* do so. The fact that biology had made women less able to fight than men was, in the opinion of many writers, a significant moral consideration why they should not be granted political equality with men. Just as some opponents of the Equal Rights Amendment in America today argue that it would be unfair to give women equal rights, since unlike men they do not risk their lives in combat, so nineteenth-century writers argued that since women were biologically unfit for fighting they should not be given the vote. Spencer had already asserted in 1876 that 'giving to man and women equal amounts of power, while the political responsibilities entailed by war fell on men only, would involve a serious inequality . . . the desired equality is therefore impracticable while wars continue' (Spencer 1898 : 757).

In the following years, women's lesser physical strength was repeatedly advanced as a reason why they *should* not, rather than as a reason why they *could* not, have equal political status with men. The 'propriety of female suffrage' is, said the American naturalist and palaeontologist, Edward Cope, 'out of the question', since woman is 'physically incapable of carrying into execution any law she may enact' (Cope 1888 : 725–26). Similarly the English bacteriologist Sir Almroth Wright, writing at the height of the suffrage demonstrations in England, argued that women should not have the parliamentary franchise because they 'cannot effectively back up their votes by force' (Wright 1912 : 8). This was despite the fact that he was fully aware that even as he wrote, the suffragettes were backing up their demand for the vote with a considerable show of physical force.

Today, the opponents of equal rights legislation occasionally bring up the biological fact of sex differences in physical strength to bolster their case. Sam Ervin, for instance, more noted for his chairmanship of the Watergate hearings than for his attitudes to women's rights, is on record as having opposed the Equal Rights Amendment on the ground that 'When He created them, God made physiological and functional differences between men and women. These differences confer upon men a greater capacity to perform arduous and hazardous physical tasks' (U.S. Senate 1972 : 9517). There is also some evidence that young children today regard male authority as legitimated by the fact that men are physically stronger and bigger than women (Ullian 1976). Among adults, however, male dominance is more often explained as the effect of biologically given sex differences in aggression rather than in terms of sex differences in physical strength *per se*. It is this explanation of male dominance that I now wish to consider.

On aggression and male dominance

The argument that biology has rendered men more aggressive than women, and that patriarchy is given by this biological sex difference, preceded the publication of E. O. Wilson's *Sociobiology* in 1975. It was, however, an argument that was readily incorporated by Wilson within the framework of his sociobiological 'synthesis'.

Sociobiology proposes, as we have seen, that various human behaviour patterns are genetically determined, that they have evolved through natural and sexual selection, and that they are the result of genetic self-interest. These propositions rest, in turn, on what the anthropologist Marshall Sahlins refers to as 'vulgar sociobiology' (Sahlins 1976 : 3), and on what the biologist Stephen Jay Gould describes as 'crude biological determinism' (Gould 1977 : 238). They depend on the thesis that 'human social organization represents natural human dispositions' (Sahlins 1976 : 4) – a thesis forwarded in the 1960s by writers like Lorenz, Morris, and Ardrey. Although many self-professed sociobiologists (e.g. Wilson 1975b; Trivers 1972) take issue with some of the claims of their 'vulgar' predecessors, they nevertheless make frequent (though not always explicit) use of their predecessors' ideas.

One of the claims of 'vulgar sociobiology' was that human aggression is instinctive and determines the shape of society. In particular, it claimed that innate aggression determined the hierarchical organization of social life and the occurrence of warfare. Subsequently, as women's rights became increasingly discussed, so vulgar sociobiology was expanded to take account of the woman question. Specifically, it was suggested that men had a greater innate tendency towards aggression and that this explained their dominance of political life.

Lorenz's earlier claim that human aggression is innate did not simply emerge out of developments internal to ethology. His thesis derived its energy and appeal from the fact that it was stated at a time of growing international tension. Following the experience of the First World War, Freud (1920a) had reversed his previous theory that human aggression was the effect of environmental factors (of the thwarting of libido). Similarly, Lorenz's challenge to environmentalist accounts of aggression in terms of frustration (e.g. Dollard *et al.* 1939) was issued during a period of mounting world tension at the time of the Berlin Wall and Cuban missile crises. Human aggression, said Lorenz, was instinctive not learned. It had evolved through competition between species for territory, through sexual selection, and through the advantages it conferred on species in providing protection to the young and in leading to the development of dominance hierarchies whereby a 'few wise males . . . acquire the authority essential for making and carrying out decisions for the good of the community' (Lorenz 1969 : 44).

Lorenz supported this last contention by reference to Washburn's and DeVore's description of male dominance in baboon societies. His concern was to use this description of baboon social life in order to support his thesis that the hierarchical organization of human society is the effect of innate aggression and ultimately of evolution. He was not concerned to investigate the question of whether biology had dictated male as opposed to female dominance in baboon or in human societies. Lorenz's successors, however, often used this same baboon data to argue that male dominance in both types of society is the effect of biologically given sex differences in aggression, and they have used films of these baboon societies to bolster their case for the evolutionary determinants of sexual divisions in contemporary human society (Chasin 1977).

Lorenz himself had argued that there were no consistent differences between the sexes as regards aggression. Indeed, he maintained that in so far as the evolution of biologically based aggressiveness had been fostered by the needs of 'brood defence', so females, including human females, were often more aggressive than males because they were the sex that more often cared for the brood: 'In many Gallinaceous birds, only the females tend the brood, and these are often far more aggressive than the males. The same thing is said to be true of human beings' (Lorenz 1969 : 33. See also Ardrey 1971). When later writers came to extend Lorenz's claim that innate aggression determines the character of human society to explain sexual divisions within society, they argued, unlike Lorenz, that biology had rendered males naturally more aggressive than females. On this basis they concluded that male domination in human society is determined by men's biologically given greater capacity for aggression; that, in the words of Steven Goldberg, patriarchy is therefore 'inevitable'.

As we have seen, Lorenz argued on the basis of Washburn's and DeVore's observations of baboon societies that social hierarchies among both non-human and human primates are determined by innate aggression. Lionel Tiger, writing in the Unesco journal *Impact of Science on Society*, applied the same data to the issue of sexual discrimination. Using Washburn's and DeVore's claim that male dominance in baboon societies is the effect of male bonding and of 'fighting ability', and ignoring their qualification that human (unlike baboon) social organization is the effect of 'custom' not of 'physical dominance' (Washburn and DeVore 1961 : 70, 71), Tiger concluded that patterns of male dominance in human as in primate societies must therefore 'originate in our genetic codes' (Tiger 1970 : 36). The next year he and Robin Fox went on to assert, again on the basis of the baboon data, that among humans 'the central arena of politics was male because of differences in male and female potentials for successful large-scale competitive bonding' (Tiger and Fox 1974 : 175–76), differences that they asserted were specified by biology.

Others, like Lorenz and unlike Tiger and Fox, have argued that male dominance in baboon societies is determined by innate aggression rather than by innate tendencies towards 'competitive bonding'. Like Tiger and Fox, however, they also use this primate data to argue for the biological determinants of human sex roles. Gray and Buffery, for instance, support their argument for the genetic determination of 'the generally greater role played by [human] males in competitive social interaction' by reference to data showing that among ground-living primates 'vigilance and defensive behaviour' is often the province of the male (Gray and Buffery 1971 : 95, 107). Similarly, Robert Claiborne uses the data documented by Sarel Eimerl and Irven DeVore (1965) in their glossy Time-Life book, *The Primates*, as evidence in support of his claim that male dominance in human society is an inheritance of our primate past. The social organization of grassland baboons is, he says, particularly relevant to the human situation because 'the grassland has been the normal habitat of our ancestors, or most of them, for more than five million years' (Claiborne 1974 : 88). He goes on to assert that male dominance in both baboon and human societies is due essentially to the effects of male hormones on behaviour; and he supports this assertion by appeal to the fact that among male monkeys aggressive and/or dominant males tend to have higher testosterone levels in the blood than other males. Since male hormones also make men more aggressive men will therefore always tend to dominate women, other things being equal.

Some have objected to this kind of biological determinist account of male dominance in the same kind of terms in which Darwin's opponents objected to his theory of evolution. Just as Bishop Wilberforce ridiculed Darwin's supporter, T. H. Huxley, by asking whether he was related by his grandfather's or his grandmother's side to an ape (Darwin 1888) – that is, by ridiculing the suggestion that there is any continuity between animal and human life – so some of today's feminist sympathizers attempt to ridicule the kind of argument put forward by Tiger and Claiborne. They ask, for instance, 'Do we really want to take the Hamadrayas baboon as our social and ethical model?' (Gross 1974: 71).

Given the way in which data on animal behaviour has frequently been used to bolster anti-feminist opinion, it is understandable that feminist sympathizers should be wary of arguments about women's destiny that are based on premises about animal behaviour. Nevertheless, Darwin did clearly demonstrate the evolutionary continuity between infra-human and human behaviour. Data on animal behaviour should not, therefore, be dismissed out of hand. It does have a bearing on the explanation of human behaviour.

What does have to be examined are the specific ways in which animal behaviour is supposed to be relevant to human behaviour. When we look at the way in which writers like Claiborne use data on baboon societies to make pronouncements about the biological determinants of male dominance in human societies, we find that his conclusions are simply not warranted by the animal data. This is not because this data does not constitute a valid model for human behaviour. Claiborne's conclusions are, rather, to be rejected on the very scientific grounds that he claims to adduce in their favour. Examination of dominance behaviour among a variety of baboon societies indicates that this behaviour is learned, not biologically determined, that therefore it does not constitute grounds for claiming that male dominance in human societies is given by biology. Comparison of various baboon societies indicates that dominance hierarchies among these primates is a learned response to situations in which space, cover, and food are limited (Bleier 1976).

Dominance hierarchies appear where foraging for food is problematic and complicated by the presence of many predators, but do not appear in habitats where food is more plentiful and predators less of a menace (Leibowitz 1975; Donelson and Gullahorn 1977). If, as this data indicates, dominance behaviour is a learned phenomenon among non-human primates, then it is more than likely also to be learned among humans – that is, a response to environmental and social factors rather than mechanistically determined by biology.

Before considering the claims of other writers as to the biological determinants of male aggression and dominance it should be pointed out that there is a further flaw in Claiborne's argument. He argues that since dominance behaviour is correlated with levels of the male hormone testosterone, this hormone must cause male dominance and that, men's domination of women is therefore the effect of their hormones. But, in fact, this correlation might well be the effect of dominance behaviour on hormone levels rather than the effect of hormone levels on dominance behaviour.

Steven Goldberg, in his notorious book *The Inevitability of Patriarchy*, advances many of the same arguments as Claiborne, and he too cites data on the effects of male hormones on human behaviour to argue his case that patriarchy (i.e. a 'system of organisation . . . in which the overwhelming number of upper positions in hierarchies are occupied by males' – Goldberg 1977:25) is biologically determined and hence inevitable. It has been found, says Goldberg, that fetally androgenized girls (i.e. girls who, in his terms, were submitted to 'male hormonalization' in utero) 'demonstrated a greater interest in a career and a lesser interest in marriage, and showed a preference for "male" toys like guns and little interest in "female" toys like dolls' (Goldberg 1977:85).

In fact, however, as E. O. Wilson (1978) points out, this behaviour might well have been due to the cortisone treatment the girls were receiving for their condition rather than to androgens operating before birth. Second, the behavioural evidence on these girls is based on their mothers' description of their behaviour, descriptions which may well have been biased by their knowledge of their daughters' medical condition, and by attitudes engendered in these mothers by the fact that their daughters had somewhat male-like genitalia at birth. Lastly, the investigators who Goldberg cites only claim that these girls showed 'a greater interest in a career'. Career interest in childhood does not, however, necessarily lead to achievement of 'upper positions in hierarchies' in adulthood. But Goldberg has to assume that it does if he is to use this data on fetally androgenized girls, as he tries to, in support of his thesis that patriarchy is determined by 'male hormonalization'.

The empirical evidence for the determination of dominance behaviour by factors in male biology is less than compelling, but this does not worry Goldberg. He claims that the universality of patriarchy *logically* entails the existence of biologically based sex differences in dominance behaviour, and in the psychological capacity for dominance. This conclusion, in his view, does not depend for its validity on the adequacy of currently available data on the hypothesized biological and hormonal causes of this behavioural and psychological sex difference. However, even if patriarchy is universal, it does not logically follow that biology has made men *psychologically* more fitted than women for high status positions in society. And without adequate independent evidence that biology has indeed provided men, more than women, with the psychological characteristics relevant to such positions, Goldberg's argument is reduced either to tautology (i.e. to the argument that men are dominant because they are dominant), or to circularity (i.e. to the argument that patriarchy must be the result of biologically determined psychological sex differences, therefore biologically determined psychological sex differences determine the existence of patriarchy).[1]

Viewed in this way Goldberg's argument thus appears absurd and hardly worth taking seriously. Unfortunately, however, his argument is given widespread credence and has been cited as authoritative by at least one women's studies text (i.e. Frieze *et al.* 1978). The same kind of argument has, moreover, been advanced in a much more influential fashion by E. O. Wilson, whose ideas receive the seal of academic respectability implied by the imprint of Harvard University Press. Like Goldberg, Wilson argues that in all existing human societies, at least among all hunting-and-gathering societies, males are dominant over females. Since, he says, this same pattern is also 'widespread although not universal' (Wilson 1975a: 568) among non-human primates, it must have characterized the society of early man. Although he argues for scientific rigour when extrapolating from animal to human behaviour – that one should only conclude that early man showed a certain trait, and that that trait 'had been subject to relatively little evolution' (Wilson 1975b: 48), if it is fairly constant within non-human primate groups – his concern to demonstrate the evolutionary roots of male dominance in contemporary society is apparently such that he is prepared to relax this criterion where this trait is concerned. Wilson acknowledges that dominance is variable among non-human primates. Nevertheless he concludes, on the basis of the primate data on this trait, that 'primitive man' must have 'lived in small territorial groups, within which males were dominant over females' (Wilson 1975a: 567).

Like Claiborne and Goldberg, Wilson also claims that male dominance in contemporary human societies is determined by men's biologically rooted greater propensity for aggression. In a newspaper article promoting sales of *Sociobiology* Wilson summarized the recent conclusions of two psychologists, Eleanor Maccoby and Carol Jacklin, about psychological sex differences as follows: 'boys consistently show more mathematical and less verbal ability than girls on the average, and they are more aggressive from the first hours of social play at age 2 to manhood' (Wilson 1975b: 50). On this basis he concluded that 'even with identical education and equal access to all professions, men are likely to continue to play a disproportionate role in political life, business, and science' (Wilson 1975b: 50).

It is not clear how men's lesser verbal ability is supposed to explain their domination of 'political life'. Indeed, one would have thought that if biology does determine social roles, as Wilson says it does, and if girls' verbal superiority is biologically determined, then women should predominate in a sphere that relies so heavily on verbal skill. The fact that women do not predominate in politics suggests that social rather than biological or psychological factors are important here. Nor is it clear how men's greater mathematical ability explains their 'disproportionate role in political life'. One is therefore left with the conclusion that as far as this aspect of male dominance is concerned, Wilson sees its justification as lying in the fact that boys are more aggressive than girls. That this is, indeed, what Wilson wishes to imply is confirmed by the fact that in his recent discussion of the biological roots of male dominance, he only cites Maccoby's and Jacklin's claims regarding sex differences in aggression, not their claims about other psychological sex differences. He now describes Maccoby and Jacklin as having shown that this sex difference is 'deeply rooted and could have a genetic origin' (Wilson 1978: 130). In sum, although Wilson (1975b) rejects Lorenz's argument that human social institutions can be reduced to innate aggression, he himself seeks to reduce at least one social institution, namely sexual inequality, to innate aggression and to sex differences in it.

Wilson does not give the details of the empirical data on which Maccoby and Jacklin base their conclusion about sex differences in aggression. This data is as follows:

1 Fetally androgenized female monkeys show male-like play patterns, including elevated levels of 'rough and tumble play', and fetally androgenized girls initiate fighting more often than their normal sisters.
2 Postnatal injection of testosterone into young female monkeys increases the frequency of 'aggression' and 'dominance' shown by them.
3 A study by Kreutz and Rose (1972) of men in prison showed that the men with the higher testosterone levels had committed the more violent and aggressive crimes. (It is noteworthy that, in the interests of arriving at this conclusion, the authors of this study included as indices of violent and aggressive crime not only acts of armed robbery and murder but also escape from institutions!)[2]

As I have shown elsewhere (Sayers 1980a), the English psychologist Corinne Hutt advances similar evidence in support of her claim that biology has made men more psychologically suited than women for occupational life. In putting forward this kind of argument Wilson and Hutt implicitly assume a continuity between the above-mentioned behaviours – 'rough and tumble play', 'aggression', 'dominance', and 'violent and aggressive crimes' – and the kinds of behaviour required in political and occupational life!

This kind of assumption is not new within biological-determinist justifications of male dominance in occupational roles. In the last century, writers also attempted to justify this dominance by appeal to the aggressive behaviour of males within the animal kingdom. Then the political theorist, Walter Bagehot, wrote: 'Take what species we like we find the males bolder, more pugnacious and quarrelsome, more adventurous and restless, and less tractable and docile'. And, as we have seen (Chapter 3 above), he claimed that these characteristics had been evolved to fit males for the role of 'defender' and 'provider'. This male role was, he said, essentially uniform throughout the animal kingdom:

> 'The man who brings home to his wife his weekly earnings, his professional fees, or his share of the profits of a business, merely

repeats on a higher scale the action of the lion who carries a deer or an antelope to his den.' (Bagehot 1879 : 208)

This constitutes a nice example of how writers on the woman question have often falsely projected human sex roles on to animal behaviour in order to claim the biological determination of those roles. It may have been true that in most middle-class Victorian families men were the sole breadwinners, but it is not true that males are the 'breadwinners' in lion families. In fact, it is the lioness, rather than the lion, who does the 'breadwinning' in the sense of trapping animal prey for her family.

E. O. Wilson is certainly better informed than was Bagehot on this matter. He recognizes that sex roles do vary in the animal kingdom, and that among lions it is the female who does the hunting. However he, like Bagehot, cavalierly assumes the functional homology of analogous behaviours in quite different environmental settings in order to argue for their common biological roots. Both writers equate hunting – whether in animal or in human societies – with wage-earning activity. Bagehot and Wilson both equate the trophies of the hunt with the wage packet of the industrial worker. Thus Wilson states:

'The building block of nearly all human societies is the nuclear family. The populace of an American industrial city, no less than a band of hunter-gatherers in the Australian desert, is organised around this unit. In both cases . . . the women and children remain in the residential area while the men forage for game or its symbolic equivalent in the form of barter or money.' (Wilson 1975a : 553. See also Ardrey 1976)

The reason these writers equate such diverse activities – foraging for game and wage earning – is that they want to argue that they are functionally homologous activities and therefore have a common biological basis. Wilson uses this equation to argue for the genetic determination of human sex roles (see p. 29 above). In this Wilson is very much in the tradition of 'vulgar sociobiology' which also bases its claims for the biological determinants of human behaviour on the assumption that primate societies can be validly equated with contemporary human societies. In Lorenz this thesis is often merely implicit; without any ado he simply redescribes Washburn and DeVore's baboon data in his own anthropomorphic terms. Whereas

Washburn and DeVore describe baboon social organization as involving male dominance, Lorenz tells his readers that in the societies studied by them the baboons were controlled by a male 'senate' (Lorenz 1969:43). Desmond Morris on the other hand, provides an explicit and virtual dictionary of purported equivalences between primate and human behaviours:

> 'Behind the facade of modern city life there is the same old naked ape. Only the names have been changed: for "hunting" read "working", for "hunting ground" read "place of business", for "home base" read "house", for "pair-bond" read "marriage", for "mate" read "wife", and so on.' (Morris 1967:84)

The major equivalence on which writers like Claiborne, Goldberg, and Wilson rest their case for the biological roots of patriarchy is that of 'dominance'. They claim, explicitly or implicitly, that male dominance behaviour is analogous, essentially constant, and therefore essentially determined by the same innate factors in apes, in hunting and gathering societies and in western industrial societies, and that in all these different types of society male dominance must therefore be essentially determined by one and the same biological cause. But, as Sahlins (1976) points out, any particular social behaviour might realize any number of different needs. The fact that activities in different social settings are all functionally similar, in that they all serve the function of dominance, does not on its own prove that these behaviours are determined by one and the same behavioural trait such as aggressiveness. But this is just what sociobiology has to assume in advancing these analogies in support of its thesis that male dominance is biologically determined.

Goldberg trades on the apparent functional similarity between 'status-seeking' behaviour in human society and 'dominance' behaviour in primate societies to argue for the common biological roots of both behaviours. Such reasoning implies that the analogy between dominance and status-seeking behaviour is symmetrical, that dominance behaviour in primates could just as well be described as 'status-seeking' behaviour and vice versa. However, as Pilbeam points out, it is quite inaccurate to use the term 'status-seeking' to describe primate behaviour, for 'status involves prestige, and prestige presupposes values – arbitrary rules or norms. That sort of behavior is cultural, human, and practically unique' (Pilbeam 1973:117–18).

Nor is it accurate to describe social organization in human

societies as universally involving male dominance behaviour. Leacock (1975), for instance, presents evidence to show that such a description of 'tribal decision making' in order to prove the universality of male dominance relies on the false projection of our social structure on to non-industrial societies. The historical data show, she says, that in many pre-class and pre-colonial societies, decision making was not hierarchically organized but was widely dispersed among the mature and elder women and men of the society. Male dominance in many of these societies today is, she claims, often the product of colonial influence. The assertion that patriarchy is universal and eternal is, therefore, historically invalid and rests on the false projection of ethnocentric conceptions of society on to primitive social organization.

Many of the writers considered above thus rely on false analogies between the social organization of primate societies and that of pre-colonial and industrial societies in order to claim that this social organization is essentially constant as far as male dominance is concerned and that male dominance must therefore be biologically determined. They also have to assume that dominance behaviour – whether it is that of guiding troop movements (in baboon societies), wage earning, occupying high-status occupations, or political activity – is essentially aggressive in nature. They have to assume this in order to argue that male dominance is dictated by men's greater biological propensity for aggression. Bagehot adduced the universality of male 'pugnacity', Hutt the fact of male 'competitiveness' and 'assertiveness', Tiger and Fox the existence of 'male competitive bonding', and Wilson that of male 'aggression' to argue for the biological roots of men's preponderance in occupational life.

In fact, however, the evidence for a linkage between dominance and aggression even in baboon societies is far from clear. Indeed it is reported (Pilbeam 1973) that the dominant male in one troop – as measured by the standard ethological criteria of frequency of completed successful matings, and influence on troop movements – was far less aggressive and, indeed, frequently lost fights with a younger and more vigorous adult male.

It is also not at all clear that human dominance behaviour, at least as measured by occupational success, is dependent on aggression and related psychological characteristics. In the absence of data proving such a linkage, writers like Bagehot and Wilson implicitly appeal to prevalent attitudes about the nature of occupational work; to the attitude that it does necessarily entail aggressive and assertive competitiveness; to an attitude shared by those in the women's movement who believe that women can only hope to achieve sexual equality through 'assertiveness training' (e.g. Osborn and Harris 1975). But this attitude is itself the product of a historically specific form of social organization, of a 'competitive society' (Klein 1971 : 87). It is the ideology of 'the bourgeois economic doctrine of competition' (Engels 1875 : 198), an ideology which, even though it was promulgated over three hundred years ago by Hobbes, is not eternally valid. Certainly, in our society, occupational life is competitive and this means that it is pure idealism and wishful thinking to hope, as some do (e.g. Gross 1974; Boulding 1977), that women's liberation will come about simply through the assertion of non-aggressive, 'feminine' virtues – virtues of non-aggression and of nurturance. Nevertheless, consideration of the actual nature of occupational work – both in contemporary industrial society and in pre-capitalist societies – indicates that it is clearly not always and essentially competitive and aggressive in nature.

The argument that male dominance is rooted in a biologically given propensity of males for aggression, like the argument that patriarchy is determined by men's greater strength and capacity for violence (i.e. the argument considered at the beginning of this chapter), is vitiated by consideration both of history and of ethnography. Such consideration shows that social organization is not universally hierarchical in form, and that occupational life is not universally characterized by aggressive competitiveness. Thus the biological determinist claim that male dominance is essentially the effect of male aggression, relying as it does on the supposed universality of male-dominated hierarchies and aggressiveness of occupational life, collapses. The evidence suggests, rather, that as in baboon societies, so in human societies male dominance is a learned phenomenon, a response to the material conditions of life; conditions that vary both historically and cross-culturally.

Notes

1 For a more extended discussion of the circularity of biological determinist accounts of human sex roles see Sayers (1980b).

2 This is pointed out by Chasin (1977). For a more comprehensive critique of this and other aspects of Maccoby's and Jacklin's empirical and theoretical arguments for the biological roots of sex differences in aggression see Tieger (1980).

Six

Sex differences in the brain

I shall now examine one last way in which attempts have been made to defend existing inequalities between the sexes in terms of biology. I shall focus in this chapter on the argument that sexual inequality is, in part at least, determined by sex differences in the brain. My reason for examining this argument is twofold. First, it is an argument that has become increasingly influential in recent years. Indeed one writer suggests that it may well constitute 'the popular version of "biology is woman's destiny" for the next few years, as sex differences in aggression have been for the last few' (Lambert 1978 : 10). Second, the history of this argument nicely illustrates the way in which biological arguments about sex roles have been influenced primarily by social factors rather than by factors solely internal to the science of biology. I shall argue this point by reference to the rise and fall of the nineteenth-century claim that sex differences in brain size determine sex differences in intellectual achievement, and then by reference to the emergence of the recent claim that sex differences in the organization of the brain determine inequalities in professional achievement.

Sex differences in brain size

The fact that women's brains are, on average, lighter and smaller than men's was repeatedly advanced in the late nineteenth century as a reason for criticizing the then current feminist campaigns for improvements in women's education. This fact was not, however, a new discovery, though the specific way in which it was then advanced in justification of sexual inequality was new. Over two thousand years earlier Aristotle (1913: 809b) had noted that women's brains were smaller than men's and had argued on this basis that they were therefore less intelligent and naturally inferior to men. Late nineteenth-century writers advanced a similar argument but claimed, in addition, that sex differences in brain size meant that it would be entirely wrong to try to give women an equal education with men.

The emergence of this thesis in the latter half of the nineteenth century is to be traced to social developments occurring in the 1860s rather than to purely scientific developments. Its foundations, however, were laid in scientific claims made at the beginning of the century, specifically by the short-lived but, at the time, influential discipline of phrenology. Many of the tenets of phrenology had been abandoned by the middle years of the century. Nevertheless, its claim that mental functions could be assessed by measurement of the shape and size of the head and skull had been retained (Young 1970).

This claim laid the basis for the subsequent argument that, since women's skulls were smaller than men's, women must therefore be less intelligent. Though the sciences of phrenology, and of its successor craniometry, thus laid the foundations for the emergence of this thesis, it was not in fact advanced until later in the century when social developments seemed to call for its assertion.

In the meantime, different social issues from those of women's rights were exercising the minds of craniometrists. These were the issues first of slavery and then of colonialism. The phrenological tenet that brain size indicated mental ability, and the discovery that non-European races generally had smaller brains than their European counterparts, raised the possibility of putting an old justification for slavery and colonialism on a firm biological footing. The enslaved and colonized races, it had been said, were naturally inferior to their masters and therefore benefitted from being subjugated to them. Craniometry seemed to offer a way of bolstering this thesis. Europeans had larger brains, and if craniometric theory was to be believed, this therefore proved their mental superiority to non-Europeans. Furthermore craniometry seemed to its practitioners to offer a way of investigating how these races might be best colonized. One noted craniometrist of the time, for instance, asserted:

> 'Civilised peoples are everywhere taking the place of savage races, or substituting for them races less warlike in character. To this end governments have to choose between two courses of action, either to destroy or to bring them together. The former, in spite of certain recent examples, is not admissible; the latter is only realisable by understanding the distinctive character of the vanquished nation, its capabilities, and the nature of its race.' (Topinard 1894:11)

Data on sex differences in the size and weight of the brain and skull were collected as part of this project of investigating the character of the 'vanquished', or colonized races. This data was not collected in the earlier part of the century either to investigate or to develop theories about the cause of sexual inequalities in society. Rather it was collected for the political purpose of defending or criticizing slavery and colonialism (e.g. Tiedmann 1836), and for the medical purposes of developing normative data on the physical dimensions of the human body (e.g. Peacock 1846; Solly 1847; Boyd 1861), and of investigating the effect of disease on the brain (e.g. Parchappe 1848). However, those collecting this data often reported it in a way that was derogatory towards women. Thus, for instance, when it was found that an intelligent woman suicide had a large brain it was argued (Topinard 1885:562) that large brains were a liability in women, not in men, because women were 'weak, impulsive, and hysterical'. Similarly, when it was found that women's brains were larger proportional to their body weight than men's this was taken as evidence not of women's superiority in this respect, but of their inferiority. It was regarded as indicative of the supposedly childish nature of women's physiology. We repeatedly find the early craniometrists (e.g. Vogt 1864; Cleland 1870) dismissing women and their brains as infantile.

Although they were thus disparaging about women, craniometrists did not, in the earlier part of the century, use the data on sex differences in brain weight to investigate the justice of existing sexual inequalities in society. They did, however, use their authority as scientists, though not their scientific data, to criticize the women's movement. Paul Broca (1868) for instance, the acknowledged leader of craniometry, criticized the feminists of his day for seeking to induce 'a perturbance in the evolution of races' (Broca 1868:50). Although he believed that women's smaller brains made them less intelligent than men (Gould 1978), he did not advance this belief in support of his anti-feminism.

In the following decades, however, the earlier craniometric data on sex differences in brain size were increasingly used both by craniometrists and by others to try to explain and justify sexual inequality. The reason this data now came to be used in this way is to be found not in developments occurring within the science of craniometry but in social changes which were then affecting middle-class women, changes that had led to vigorous campaigns aimed at remedying existing sexual inequalities in society.

Economic and demographic changes during the second quarter of the nineteenth century had resulted in a large increase in the number of single middle-class women needing employment.[1] This had led in turn to increased demands by middle-class women for greater access to education and employment, and for the vote so that they could give legislative force to these demands. We thus find a woman's suffrage amendment to the Representation of the People Bill being proposed in 1867 – an amendment which, despite the fact that the Muncipal Franchise was granted to women in 1869 was to be only the first of many unsuccessful attempts to secure women the parliamentary vote in England. Also in 1869, *The Subjection of Women* was published. In it John Stuart Mill argued that women's education should be improved and that their lack of achievement in the past had been due to the inferior quality of their education.

It was these social developments rather than developments internal to the science of craniometry itself that led to its data now being advanced for the first time as reasons why women should not be granted equal education with men. Writers of this period had already commented on sex differences in intellectual style (e.g. Buckle 1858). Now these differences and the apparently associated differences in brain size were cited again and again as justification of existing sexual inequalities in education. The argument of one James McGrigor Allan, an anthropologist, is typical in this respect. On the basis of the earlier craniometric data on sex differences in brain size, and of its claims about the infantile quality of the female brain, Allan argued that women's lack of intellectual achievement was determined by the nature of their brains, not by the nature of their education. On this basis, he claimed that 'any attempt to revolutionise the education and status of women on the assumption of an imaginary sexual equality, would be at variance with the normal order of things' (Allan 1869:213).

The use of craniometric data in arguments about social issues was not, as we have seen, new to anthropology and craniometry. Such data had for several decades been used to address the issues of slavery and colonialism. Its use by anthropologists for the purpose of addressing the woman question was new however, and reflected two social, as opposed to scientific, developments: first, that this question had become increasingly pressing during the 1860s; and, second, that the professional anthropology societies were concerned to make their discipline relevant to contemporary political questions (Fee 1979). These societies were therefore happy to give room in their journals to the discussion of women's rights since this was an issue that was now receiving prominence in public debate.

Other journals were, if anything, even more willing to give space in their columns to the discussion of the woman question in terms of craniometric data. Thus, we find reports and articles on this topic in the applied journals (e.g. *Medical Record*, *Educational Review*), in journals designed to acquaint the general public with the recent discoveries of science (e.g. *Popular Science Monthly*), and in journals of more general interest (e.g. *Nineteenth Century*, *New Review*). Reputable scientists and political theorists, doctors, and others contributed articles to these journals, arguing on the basis of the craniometric data that women were innately inferior to men, and that this inferiority could not be significantly reduced by educational reform. Why, they asked (e.g. Van de Warker 1875; Sergi 1892), were there so few women scientists and artists? – Not, they answered, because women's education had been inferior to men's but because women's small brains had rendered them incapable of high intellectual achievement.

In recent times, it has been repeatedly suggested that craniometry seemed admirably suited to the anti-feminist cause of the late nineteenth century (e.g. Haller and Haller 1974; Shields 1975; Gould 1978). Less frequently noted, however, is the fact that some of the tenets of craniometry were often viewed at the time as more in accord with feminist than with anti-feminist claims. Ironically, although feminists have frequently been accused of ignoring scientific and biological theory, in this instance, they were arguably more faithful to it – at least to craniometric theory – than were the opponents of feminism.

Broca (e.g. 1862, 1873) insisted that brain size was affected by the exercise of the intelligence. Accordingly both he (e.g. Broca 1879) and his pupil, Topinard, concluded that sex differences in brain size were as much the effect as the cause of the differences in the opportunities given to men and women to exercise their intelligence. Topinard, for instance, maintained that:

> 'The reason that the brain of woman is lighter than that of man is that she has less cerebral activity to exercise in her sphere of duty. In former times it was relatively larger in the department of Lozère, because there the woman and the man mutually shared the burden of their daily labour. The truth is, that the weight of the brain increases with the use which we make of that organ.' (Topinard 1894: 121)

Those sympathizing with feminism were happy to appropriate this craniometric thesis for their cause. If, they argued (e.g. Distant 1874; Gardener 1887), sex differences in brain size were the effect, not the cause, of sexual inequalities in social life, then this was all the more reason why these inequalities should be redressed. Similarly, others (e.g. Buchner 1893) argued that if, as the craniometrists (e.g. Le Bon 1878) had shown, sex differences in brain size increased with 'civilization', and if this increase was not the product of evolution, as some (e.g. Bagehot 1879) had claimed, but was instead due to the fact that 'civilized' women had less cause to use their brains than 'primitive' women, then this again called into question the lot accorded to women by 'civilized' society.

Although those opposed to the feminist cause claimed to be more faithful to biological theory than the feminists, they found it difficult to do justice in their arguments against feminism to Broca's thesis about the effects of the environment on brain size. Some (e.g. Bagehot 1879) entirely ignored this craniometric thesis and argued that craniometry had demonstrated unequivocally that women's mental inferiority was determined by cranial inferiority. Others (e.g. Hammond 1887a; Romanes 1887) toyed with this thesis but, without criticizing it, went on to draw a biological determinist conclusion from the craniometric data. George Romanes, for instance, raised the possibility that sex differences in brain size might have been caused by existing sexual inequalities in society. There might, he said, be some justice in the feminist claim 'that the long course of shameful neglect to which the selfishness of man has subjected the culture of woman has necessarily left its mark upon the hereditary constitution of her mind'. Perhaps, he suggested, this 'neglect' had led to 'the missing five ounces of the female brain' (Romanes 1887: 665–66). But, like other anti-feminists, he also and simultaneously advanced the biological determinist thesis that this difference in brain weight was determined by biology, and could not 'be explained on the hypothesis that the educational advantages enjoyed either by the individual man or by the male sex generally through a long series of generations have stimulated the growth of the brain in the one sex more than in the other' (Romanes 1887: 655, n. 1).

Despite such inconsistencies in the claims of these anti-feminists about women's brains, these claims continued to be adduced as grounds for complacency about existing sexual inequalities in education, work, and politics right up to the turn of the century. Clara Zetkin, for instance, reported in 1896 that women's smaller brain was still being cited then in arguments against 'intellectual labour by women' (Zetkin 1896: 195). And an Englishman writing in 1900 reported that, 'Only last year, two public men in England . . . gave it as their reason for voting against a certain citizen claim on behalf of women that their brains are smaller than men's' (Sutherland 1900: 803). Nevertheless, within a few years of the dawning of the new century the claim that women's smaller brains had dictated their lesser achievements was dropped from the armoury of arguments against the feminist cause. And by 1929, Havelock Ellis confidently proclaimed that this argument had well and truly collapsed. Although one or two writers (e.g. Porteus and Babcock 1926) had attempted to resurrect it, their appeal had generally fallen on deaf ears.

I shall argue that the cause of the decline of this thesis in the early years of the twentieth century, like that of its first emergence in the late 1860s, is to be found in changes that were then occurring in women's social status. The decline of the claim that sexual inequality was determined by sex differences in brain size was, I shall contend, primarily due to these social factors. Others, however, have suggested that the demise of this claim was primarily due to scientific developments occurring in the early years of the century.

It has been argued, for instance, that the thesis that women's smaller brains made them the intellectual inferiors of men was abandoned in the early twentieth century because new data then came to light about the brain, data that rendered this claim scientifically insupportable. Franklin Mall's (1909) scientific demonstration that neither brain weight nor measures of the cerebral convolutions afforded a reliable basis for distinguishing between brains according to sex has been cited as a major cause of the rejection of the thesis that sex differences in brain size determine sexual inequality (e.g. Woolley 1910; Lincoln 1927; Fee 1976). As I have indicated, phrenology had given rise in the first place to the confident assertion that brain size was a reliable index of intelligence, and its leading exponent – Franz Gall – had claimed to be able to distinguish male and female brains simply on the basis of their physical appearance (Walker 1840). Now Mall had shown this claim to be false. It was for this reason, it is argued, that women's small brains were dropped from the baggage of anti-feminist polemic.

It is doubtful, however, whether Mall's data did constitute the effective cause of the demise of the thesis that the unequal achievements of the two sexes were determined by the differences in the size of their brains. Earlier proponents of this thesis (e.g. Hammond 1887b) had countenanced with equanimity the evidence that the sex of an individual brain could not be determined simply on the basis of its physical appearance. Such evidence, they said, constituted no disproof of their claim that women's brains determined their intellectual inferiority since this claim was concerned with the difference in the *average* size of men's and women's brains not with the character of *individual* brains. On the matter of the difference in the average size of men's and women's brains there was remarkably little disagreement over the facts; both advocates (e.g. Buchner 1893; Ellis 1929) and opponents of the women's cause agreed with each other that, on average, women's brains were indeed smaller than men's.

Why, then, does the psychologist Mary Brown Parlee now claim that, 'What was being disputed were the facts themselves; there was no disagreement about the interpretation of any sex differences in brain weight that might be found' (Parlee 1878 : 66)? The only major dispute regarding the facts concerned the relative size of the frontal and parietal lobes in men and women. Whereas earlier investigators had claimed that the frontal lobes were relatively larger in men (e.g. Huschke 1854; Schaaffhausen 1868), later investigators reversed this judgment and asserted that the frontal lobes were relatively smaller, and the parietal lobes relatively larger in men (e.g. Crichton-Browne 1879; Cunningham 1892). This change in the facts had, however, been readily accommodated by the biological determinists, who persisted in arguing that sex differences in brain size dictated sex differences in intelligence. Indeed, despite their avowed respect for science, these writers often displayed a remarkably cavalier attitude towards the scientific data. One writer, for instance, happily countenanced this apparent change in the facts as follows:

> 'The frontal region is not, as has been supposed, smaller in woman, but rather larger relatively . . . But the parietal lobe is somewhat smaller. It is now believed, however, that a preponderance of the frontal region does not imply intellectual superiority, as was formerly supposed, but that the parietal region is really the more important.' (Patrick 1895 : 212)

Sexist prejudice seems to have blinded these scientists to the very biological facts to which they claimed to be so faithful in their critiques of the feminists. As Havelock Ellis pointed out:

> 'it may, indeed, be said that it is only since it has become known that the frontal region of the brain is of greater relative extent in the Ape than it is in Man, and has no special connection with the higher intellectual processes, that it has become possible to recognize the fact that that region is relatively more extensive in women.' (Mosedale 1978 : 4)

Since change in the way the facts were perceived was so easily accommodated to the anti-feminist cause it could not have been the main reason for the collapse of its thesis that sexual inequality was determined by sex differences in brain size.

Parlee goes on to say that 'Today, there is no firm evidence of sex differences either in the overall size or the proportion of parts of the human brain' (Parlee 1978 : 66). Perhaps this denial is motivated by the belief that to acknowledge such a biological sex difference would be to concede the case to the opponents of feminism. In fact, the evidence of sex differences in the overall size of the brain is overwhelming and no one has ever successfully challenged Aristotle's (1912 : 652–53) contention that the brain is larger in men than in women. Nor was this biological fact in any way fatal to feminism.

Others have suggested that it was not disagreement about the facts themselves that led to the demise of the claim that sexual inequality is determined by sex differences in brain size. Instead, they argue, this claim was abandoned because scientists now changed their opinion about how these facts should be interpreted. The thesis that sexual inequality is determined by sex differences in brain size was dismissed, it is said, because scientists came to reject absolute brain size as a measure of intelligence. As one writer of the period put it:

> 'The belief that the brains of females differ from those of males has been widely accepted, and has been thought to be conclusive evidence of the permanent inferiority of the female mind . . . It is now a generally accepted belief that the smaller gross weight of the female brain has no significance other than that of the smaller average size of the female.' (Woolley 1910: 335)

This latter 'belief' had, however, been repeatedly canvassed in the preceding century but to no avail then as far as the claim that sexual inequality is determined by sex differences in brain size was concerned. Forty years earlier, for instance, Darwin had remarked that man's 'brain is absolutely larger, but whether or not proportionately to his larger body, has not, I believe, been fully ascertained' (Darwin 1896: 557). It had been regularly pointed out in those years that relative, rather than absolute brain size was probably a more valid measure of intelligence, since the absolute measure wrongly credited elephants and whales with more intelligence than humans (e.g. Gardener 1887). This so-called 'elephant problem' (Fee 1979: 421) had resurfaced again and again in discussion both of the appropriate measures to use in comparing the brains of different races (e.g. Tiedmann 1836; Distant 1874), and later in discussion of the appropriate measures to use in comparing the brains and hence the intelligence of the two sexes. As a result, a number of alternative measures to that of absolute brain size had been proposed. We find, moreover, that these measures tended to be rejected or adopted largely on the basis of the very sexist prejudices which they were later to be used to investigate. Some proposed measuring the ratio of brain to body weight but this measure was soon found to favour women (e.g. Van Soemmering 1788) and was thereafter abandoned (e.g. Gall 1835). Others proposed measuring the ratio of brain weight to body height, a measure which favoured men and was generally adopted (e.g. Parchappe 1848; Topinard 1885).

These various measures had been recommended as a way of resolving the elephant problem. The raising of this problem by writers in the early twentieth century was not, therefore, new. And just as this problem had previously been raised only to be dismissed by those (e.g. Tiedmann 1836) who claimed that absolute brain size was an adequate measure of intelligence, so these later writers (e.g. Sutherland 1900) were also often happy to recognize the problem while continuing to assert that sex differences in absolute brain size did constitute a valid index of women's intellectual inferiority.

Disillusion with absolute brain size as a measure of intelligence cannot therefore have been the primary cause of the decline of the thesis that women's smaller brain dictated their lesser achievements. On the contrary, this thesis seems to have declined for the same social reasons that had given rise to it in the first place, namely changes then taking place in women's social status. By the turn of the century economic pressures on middle-class women resulted in their increasingly having to find employment,[2] and to prove themselves the intellectual equals of men in work. Moreover, despite the opposition to reforms in women's education, many such reforms had successfully been completed by the turn of the century. As a result of women's new-found access to higher education and of their 'triumphs on the campuses' (Fee 1976 : 205)[3] it was no longer plausible to maintain, as earlier writers had, that women were less intelligent than men.[4] Nor, therefore, did it seem so reasonable as it had before to attribute women's relative lack of achievement in the past to inferior intelligence, rather than to inferior education.

In 1869, it had been asserted that 'the history of humanity is conclusive as to the mental supremacy of the male sex . . . In the domain of pure intellect it is doubtful if women have contributed one profound original idea of the slightest permanent value to the world!' (Allan 1869 : 210). Two years later Darwin had written in a somewhat similar, though more moderate vein of the mental inferiority of women:

> 'The chief distinction in the intellectual powers of the two sexes is shewn by man's attaining to a higher eminence, in whatever he takes up, than can woman . . . if men are capable of a decided pre-eminence over women in many subjects, the average mental power in man must be above that of woman.' (Darwin 1896 : 564)

So reasonable had Darwin's argument appeared in this matter that it had repeatedly been cited by subsequent writers (e.g. Delauney 1881; Romanes 1887) as authoritative grounds for concluding that women were the mental inferiors of men.

The premise of Darwin's argument, however, became increasingly implausible as women came to prove that they were capable of equalling men's 'pre-eminence'. His conclusion regarding women's intellectual inferiority therefore also became untenable. And as a result it came to seem absurd to adhere to a measure of intelligence, like that of brain size, which led to this kind of conclusion.

In 1929 Havelock Ellis criticized the various measures of the brain then being used as indices of intelligence – measures such as the ratio of brain size to body weight, and of brain size to body height – as 'not quite fair to women' (Ellis 1929 : 215). Considerations of equity had not, however, weighed with earlier writers on this subject. Indeed, so strong had been their conviction of the intellectual inequality of the two sexes and of the different races of mankind that only those measures that corroborated this 'unfair' belief were adopted. That is, grounds of in-equity rather than of equity had more usually prevailed in governing the choice of physical measures of intelligence. Brain to body-weight measures of intelligence had, as we have seen, been rejected at least partly on the sexist ground that they implied that women might be brighter than men (e.g. Sutherland 1900). Similarly cranial height had been rejected as a measure of intelligence on the racist ground that it gave the advantage to 'Kaffirs, Negroes and Australians' (Fee 1979 : 425). In like manner, cranial capacity was dismissed by the Frenchman, Jean Finot (1913), on the chauvinist ground that it implied that Eskimoes were more intelligent than Parisians!

Considerations of equity became an issue only as women came to prove themselves the equals of men, and even then these considerations were not extended to individuals of non-European races. Indeed, those arguing for a more just interpretation of the brain data in regard to women, far from extending the principle of justice to non-Europeans, often appealed to racist sentiment to support their appeals on behalf of women. For instance one writer who criticized the thesis that women's brains had rendered them innately inferior, concluded his defence of feminism with the following rhetorical flourish: 'with what feelings,' he asked, 'must a highly educated American woman view a dirty, idiotic negro shoe-black or street sweeper going to the ballot-box while she herself remains excluded from it' (Buchner 1893 : 176).

Stephanie Shields suggests that this whole 'debate concerning the importance of brain size and anatomy as indicators of intelligence' declined in part because of the 'development of mental tests' (Shields

1975 : 72). Certainly these tests indicated that men and women were intellectual equals. But this was because these tests had been constructed on the basis of the new-found assumption that women were intellectually equal to men. This belief was built into these tests, it was not a result of their use. Items were excluded from them if they discriminated between the sexes. These tests did not therefore cause so much as reflect a change in attitude towards women's intelligence. The principle of equity – of the mental equality of the sexes – was built into the IQ test. This principle was not, however, extended to members of different races or to members of different social classes. Indeed, far from ensuring that items were not included if they were unfair to, and discriminated between different races and different social classes, the IQ test was instead often standardized and validated in terms that simply transferred the racist and class bias of craniometry to psychometry. Psychometric tests of intelligence have usually been standardized on predominantly white, middle-class populations (Ryan 1972), and have been validated in terms of success within the class structure, in terms of educational and occupational success.[5] As a result, although the development of intelligence tests was accompanied by a decline in the belief that women are the intellectual inferiors of men, it in no way helped to discredit the belief that Blacks are the intellectual inferiors of Whites, or the belief that the working class are the intellectual inferiors of the middle class. Indeed these tests, far from contributing to a decline of such prejudices, have often been used to bolster them.

We have seen that the decline of the purportedly scientific claim that sexual inequality in society is determined by sex differences in brain size was due as much, if not more, to social factors – to factors external to the sciences of craniometry and of the psychometry which replaced it – than to discoveries within these sciences themselves. Nevertheless scientific research did contribute to the demise of this doctrine. By using the very criterion of intelligence (namely, that of educational success) which was to be incorporated within the IQ test and which was to prove so useful subsequently to those wishing to 'prove' the mental inferiority of the lower classes, Karl Pearson (1902) demonstrated that brain size was not an adequate measure of intelligence. He showed that intelligence in children, as measured by their teachers' estimates of their ability, did not correlate with the size of their heads. In this way he and his associates (e.g. Lee 1901) did indeed contribute to the decline of brain size as a measure of intelligence. Nevertheless the main causes of the demise, as of the initial emergence of the claim that sexual inequality in mental achievements is determined by sex differences in brain size (and in intelligence) were changes in the position of women in society.

Sex differences in brain organization

I shall now consider the recent version of the claim that sexual inequality is determined by sex differences in the brain. The character of this claim, just like that of its nineteenth-century forerunner, has been determined as much by social as by scientific factors. As a result of the historical defeat of this earlier thesis scientists (e.g. Newcombe and Ratcliff 1978) are now much more wary of concluding, on the basis of sex differences in the size of various parts of the brain, that these differences imply differences in intellectual function. The improvement in women's social status over the last century has led to changes in some of the other details of the current claim that sexual inequality is determined by sex differences in the brain. Nineteenth-century writers argued that sex differences in brain size had determined women's lack of achievement in all intellectual professions from science to literature. Today, as more and more women have demonstrated their abilities in an increasing range of different professions, so the scope of the current doctrine regarding the cortical determinants of women's lack of intellectual eminence has been narrowed. Women's 'deficient' brains, it is now said, explain their lack of achievement in specific professions – particularly in engineering and architecture – not in professional life in general. Lastly, the nineteenth-century version of this doctrine used data collected almost exclusively by men and almost always for the primary purpose of investigating issues other than that of sex differences. Nowadays, however, much of the research on sex differences in brain function is being conducted by women and, perhaps because of this, much of it has been collected with the specific aim of investigating sex differences.

The discovery that there might be differences between the sexes in the organization of their brains – the discovery which feeds the current doctrine regarding the cortical determinants of sexual inequality – was not, however, made initially as a result of investigating sex differences. Researchers had long been interested in Broca's discovery, in 1861, that the left hemisphere of the brain seemed to be the seat of language function, and in Hughlings Jackson's observation, in 1864, that patients with right hemisphere lesions 'did not know objects, persons or places' (Ornstein 1973:86). Such observations suggested that the brain was not functionally symmetrical, but that the two halves of the brain or cerebral hemispheres served different mental functions.

Since much of the subsequent research on this topic, at least in England, was done on people who had sustained brain injury during combat duty – that is, only on men – it could not reveal differences between men and women in the asymmetrical organization of their brains. It is for this reason, in part, that English researchers (e.g. Newcombe and Ratcliff 1978) who are now adding their scientific authority to the doctrine that sexual inequality is determined by cortical factors showed little interest in sex differences as long as their research (e.g. Newcombe and Ratcliff 1973) was confined to work with the war-wounded. That is, the social fact that women have been excluded from combat duty explains, in part, the earlier neglect of the possibility of sex differences in the asymmetrical organization of the brain.

The discovery that there might be sex differences in this aspect of brain function resulted from developments in neurosurgery, and the treatment of epileptic patients. In the course of studying the effects of neurosurgical operations for the relief of epilepsy it was found that patients undergoing these operations responded differently to tests of brain function according to sex. Temporal lobe surgery, for instance, when conducted on the dominant side of the brain (i.e. in the hemisphere which specializes in language function), disrupted the performance of men on verbal tests but not of women (Lansdell 1961). On the other hand, when this surgery was conducted on the non-dominant side of the brain, it disrupted performance on a test of artistic judgment, again in men but not in women (Lansdell 1962). These early discoveries, like the earlier craniometric data on sex differences in the brain, were not made as the result of investigating sex differences *per se*. These discoveries were instead incidental to research on a different, though related, topic, namely that of the general nature of brain organization in humans, and were noted at the time as merely 'surprising' epiphenomena (Lansdell 1962:854).

We have seen above that the early craniometric data on sex differences in brain size were not used to examine the causes of sexual inequality in society until these inequalities came to be vigorously criticized and challenged by feminist sympathizers in the 1860s. Similarly, the data of the early 1960s on sex differences in brain organization were not used then to explore the causes of sexual inequality in society. Again it was social, not scientific, factors that led to the attempt to explain this inequality in terms of the brain.

The emergence of a vociferous women's movement in the late 1960s and early 1970s had the effect of focusing public attention on the issue of sexual inequality, and hence on the related issue of sex differences in behaviour. As a result of this social development increasing research attention was now given to investigating the psychological differences between the sexes. Whereas these differences had previously been virtually ignored by psychologists (Lloyd 1976), they now became a major focus of psychological research. Increasingly research results came to be routinely analysed in terms of sex differences (Etaugh and Spandikow 1979), and the earlier findings of sex differences in brain organization were given greater attention. These earlier sporadic findings were now incorporated into general theories (e.g. Buffery and Gray 1972; Levy 1972) and these theories, in their turn, then generated still further research into possible sex differences in brain organization (e.g. McGlone and Davidson 1973; Witelson 1976; Waber 1976).

The conclusions drawn from this research have been somewhat contradictory with some researchers espousing one interpretation and others an entirely opposite interpretation of the data. The most plausible theory, at present, seems to be that advanced by Jerre Levy. I shall therefore limit myself here to outlining her theory about sex differences in the organization of the brain. In common with other researchers in this field, Levy was not initially interested in investigating sex differences. Indeed much of her research, like much pre-1970s psychological research, used male subjects only; it reflected the pervasive male bias of psychological research at that time.[6] She was concerned to study differences in the brain organization of left- and right-handed men. Previous research had indicated that language functions are represented unilaterally (i.e. in only one hemisphere – the left hemisphere) in right-handed men, but bilaterally (i.e. in both hemispheres) in many left-handed men. Levy (1969) found that left-handed men did relatively poorly on spatial tasks as compared to right-handed men. She suggested that this might be because left-handed, unlike right-handed, men process spatial and verbal information in the same hemispheres, and that this results in conflict between the two types of information, which leads to inferior spatial performance.

Subsequently, and I would suggest partly as a result of the prominence then being given by the women's movement to the subject of sex differences, Jerre Levy (1972) extended her theory to take account of the finding that girls perform worse on spatial tests than boys. She suggested that perhaps girls (like left-handed men) process spatial information in the same part of the brain as they process verbal information and that as a result these two types of information interfere with each other, thus disrupting performance on spatial tasks. That is, she hypothesized that the poor spatial performance of girls (like that of left-handed men) might be due to their language functions being represented bilaterally rather than unilaterally.

Levy herself has not, as far as I know, used her theory about sex differences in brain organization to explain existing sexual inequalities in the professions. Nevertheless, like the nineteenth-century craniometrists and despite her sex, her remarks about women are often disparaging. She claims that specialization of one hemisphere for language, the other for spatial information (which, she suggests, characterizes right-handed men but not women) indicates a more 'perfect' evolution of the mechanisms involved (Levy 1972 : 122). By implication, therefore, she dismisses women (along with left-handed men) as imperfectly evolved. Subsequently she has also characterized women's brains as flawed by a 'right hemisphere deficit' (Levy and Reid 1978 : 125).

Others who have reported her theory, and who have themselves done research on sex differences in brain organization, have not merely been disparaging about women's brains, but have also used this research to bolster their belief in the futility of current feminist demands for equality in professional life. Sex differences in brain organization, they say, determine sex differences in spatial ability and hence the under-representation of women in professions which use spatial ability – professions like engineering and architecture. And just as popular magazines of the nineteenth century gave prominence to the craniometric data when it seemed to be serviceable to such conservative attitudes towards sexual inequality in society, so we find popular magazines today giving prominence to this psychological research now that it appears to offer a basis for shoring up conservative opposition to current demands for equality in the professions. For example, an article in the glossy American magazine *Horizon* publicized this research and claimed on the basis of it that

> 'The fact that women have not made large inroads into fields such as engineering, in which a high degree of spatial ability is needed, is clearly due to something more fundamental than the hostility of male engineers. It is due, in part at least, to biology.' (Lamott 1977:43)

So germane do these findings about sex differences in brain organization appear to the current political debate about the justice of continuing sexual inequalities in professional life that they are now regularly singled out for coverage in newspaper reports of scientific meetings. For instance, a report of a meeting of the American Association for the Advancement of Science gave prominence to the claims of one Dr William Blackmore, who maintained that findings about sex differences in brain function indicate that sexual inequality in the professions is the effect of biological rather than of social factors. 'I simply don't believe,' he is reported as saying, 'that women have been locked out of brilliant creative accomplishments by social conditions'. Male hormones, he suggested, result in superior right brain performance in men and it is this biological factor, not 'social conditions', that dictates the greater accomplishments of men in architecture, engineering, and art (Rodgers 1977:1).

We saw above that nineteenth-century feminist sympathizers, faced with similar justifications of existing sexual inequalities in intellectual achievement, insisted that sex differences in the brain were as much the effect as the cause of these inequalities, and that therefore these biological differences constituted an indictment rather than a justification of sexual inequality. Some of those who wish to see further improvements in women's social status today likewise suggest that sex differences in the brain may be the effect rather than the cause of sexual inequality. Judith Sherman (1977), for instance, contributing to a U.S. Government advisory document on women and mathematics education, suggests that female biological development in interaction with the traditional style of female education results in girls relying excessively on using their dominant hemisphere to process both verbal and spatial information. She argues that this in turn impedes their spatial performance. Given appropriate changes in their education, she suggests, women might well come to use their right rather than their left hemispheres to process such information. Although she acknowledges that there may well be sex differences in the development and functioning of the brain, she also believes that these differences do not constitute an insuperable obstacle to sexual equality. Instead she recommends that these differences, and with them sexual inequality, can and should be reduced through the institution of appropriate reforms in women's education.

Sherman is not alone in putting forward this kind of interactionist account of sex differences in spatial ability; that is, an account which views sex differences in brain development and function both as causing, and also as being caused by, sex differences in experience. Lauren Jay Harris (1977), like Sherman, points out that girls' verbal functions mature more quickly than those of boys. He suggests that this in turn results in mothers stimulating verbal skills more in their daughters than in their sons. This experience, he claims, then leads girls to rely on their dominant hemisphere (i.e. on the hemisphere that specializes in language function) to process both linguistic and spatial information, despite the fact that the latter would be better processed by the non-dominant hemisphere.

We saw above that in order to strengthen their case against feminism, many nineteenth-century writers ignored some of the detailed claims of craniometry – the very science on which they claimed to rest their anti-feminist arguments. In particular, they often ignored or dismissed Broca's interactionist thesis that sex differences in the brain were as much the effect as the cause of inequalities in social life. In the same way, those who now argue that science has shown that sexual inequality is determined by sex differences in brain organization also neglect to mention that many of the scientists (e.g. Harris 1978; McGuinness and Pribram 1979) whose work they cite regard these differences as both affecting and being affected by differences in the experience and behaviour of the two sexes. Instead they imply that these scientists uniformly advance a biological determinist rather than an interactionist account of the relation between brain organization and spatial ability, and that they have thereby demonstrated that 'the different representation of the sexes in various professions may have a biological basis' (Goleman 1978 : 59), and that 'genuine [i.e. cortically based] sex differences in cognitive skill are reflected in choice of career and professional achievement' (Newcombe and Ratcliff 1978 : 195).

These claims rest ultimately on the fact that girls do worse than boys in tests of spatial ability. But sex differences on these tests are very small. They account, at most, for 4 per cent of the variance (Sherman 1978). They cannot, therefore, explain or justify the extremely large discrepancy between the numbers of men and women entering professions such as architecture and engineering which are said (e.g. Coltheart 1975) to depend on the kinds of skill measured by these tests.

Several psychologists (e.g. Sherman 1979; Herron *et al.* 1979) also point out that many of the behavioural measures of cerebral dominance – measures like those of handedness and of writing posture – that have been used to investigate sex differences in brain organization are not reliable as indices of dominance. Furthermore, others (e.g. Star 1979; Bryden 1979) have demonstrated that most of the psychological tests used in this research lead to ambiguous results. Sex differences on these tests could be due to sex differences in brain organization. They could alternatively be the result of men and women adopting different strategies to deal with these tests, strategies that are determined by non-cortical factors. Lastly, even if the results of these tests do indeed reflect underlying sex differences in brain organization, no one has yet convincingly demonstrated that such differences *cause* sex differences in spatial ability (Jacklin 1979).

These objections to the research on sex differences in brain organization are relatively recent. It remains to be seen whether such objections, made from within the scientific discipline that gave rise to the claim that sex differences in brain organization determine sexual inequalities in the professions, will also lead to the demise of this claim. If the history of the decline of the analogous nineteenth-century doctrine is anything to go by one would predict that this current claim will collapse only given further changes in the social status of women. It will finally collapse, I predict, only given a situation in which women prove themselves equally competent to men in professions like engineering and architecture. In order for women to be able to do this, however, it will be necessary to struggle against the discriminatory practices that currently obstruct their entry into these professions, practices which are now being justified by some, as we have seen, in terms of sex differences in brain organization.

Notes

1 The 'abnormal excess of women' (Spencer 1898 : 768) was frequently noted at the time. For a useful account of the causes of this demographic change see Banks and Banks (1965).

2 For contemporary comments on this development see, for example, Zetkin (1896), Buchner (1893), and Gwynn (1898).

3 Women's 'distinguished scholarship' (Stevenson 1881 : 167) was repeatedly noted by contemporary commentators and was, indeed, a major source of complaint of those (e.g. Cope and Kingsley 1895) who opposed improving women's rights in education.

4 Woolley was therefore probably more correct to attribute the growing disinterest in the issue of sex difference in intelligence to 'the successful competition of women in graduate work' (1910 : 341) than she had been in attributing it to the rejection of absolute brain size as a measure of intelligence.

5 For evidence that these constitute standard criteria for validating IQ tests see, for example, Annastasi (1968 : 108–10).

6 This general bias in psychological research has been noted, for instance, by Carlson and Carlson (1960), Schultz (1969), and Chetwynd (1975).

PART II

Feminist theory and biology

Seven

The social construction of female biology

I have now discussed a number of ways in which the opponents of feminism have sought to appropriate biology for their cause. Although feminists have been accused of neglecting and ignoring biology, they have in fact suggested a number of ways of accommodating the facts of biology in their analysis of women's situation. I shall now consider some of the accounts that feminists have offered of the relation between biology and women's destiny. In this chapter I shall focus on a theory that I shall term 'social constructionism'; that is, the theory that biology determines sexual divisions in society primarily via the way it is socially constructed within that society. I shall first examine a structuralist version of this theory, a version which – in one form or another – has had a fairly widespread influence on recent feminist theory. I shall then look at another version of this theory, one that has focused on social attitudes towards menstruation, and on the way these attitudes and superstitious beliefs contribute to the social subordination of women.

A structuralist version of social constructionism

Sherri Ortner's paper, 'Is female to male as nature is to culture?', constitutes a very clear statement of structuralist social constructionism. I shall therefore begin by considering her argument. Ortner starts by declaring that she wants to see 'genuine change' come about in women's position in society. On this basis she dismisses biological determinism and its implication that sex roles are relatively fixed. She back up this dismissal thus: 'Without going into a detailed refutation of this position, I think it fair to say that it has failed to be established to the satisfaction of almost anyone in academic anthropology' (Ortner 1974:71).

The main reason that 'academic anthropology' often rejects a biological determinist account of women's social status is that it rejects the premise of this account, namely that the character of women's social subordination is essentially constant in all societies. Ortner, however, retains this premise but seeks some universal other than biology to explain it: 'if we were not to accept the ideology of biological determinism, then explanation, it seemed to me, could only proceed by reference to other universals of the human cultural situation' (Ortner 1974:83). She believes she has found this universal in the supposedly uniform way in which all societies regard women, on the basis of their biology, as closer to nature than men. It is, in her view, this social attitude towards female biology, not biology itself, that explains the apparently unchanging nature of women's social subordination.

The reason, she suggests, why societies have always construed women's biology as placing them closer to nature is because of the seemingly greater involvement of women's biology in the reproduction of the species: 'Woman's body and its functions, more involved more of the time with 'species life', *seem* to place her closer to nature, in contrast to man's physiology' (Ortner 1974:73. My emphasis). Ortner finds support for this thesis in de Beauvoir's claim that menstruation, pregnancy, and childbirth result in the female being 'more enslaved to the species than the male, her animality is more manifest' (Ortner 1974:74). But whereas de Beauvoir claims that woman's biology *actually* renders her 'more enslaved to the species', Ortner claims that it only *seems* to involve her more with 'species life', that it is not biology *per se*, but the social construction of it, that places women closer to nature.

The social subordination of women is due, she suggests, to the fact that all societies value culture above nature. In this social system of valuation women are regarded as inferior to men because their biology makes them appear to be closer to nature. Societies universally value culture above nature, she says, because they are all

'engaged in the process of generating and sustaining systems of meaningful forms (symbols, artifacts, etc.) by means of which

> humanity transcends the givens of natural existence . . . We may thus broadly equate culture with the notion of human consciousness (i.e. systems of thought and technology), by means of which humanity attempts to assert control over nature.' (Ortner 1974:72)

In generating social and cultural life, human groups, she suggests, necessarily have to conceptualize themselves as distinct from, and as transcending the givens of nature. Within this universal system of social classification, she says, women are always perceived as inferior because their biology is construed as placing them closer to the very nature that it is the business of society to transcend.

Ortner thus arrives at an essentially idealist explanation of women's social subordination: women are subordinate, in her view, because of the ideas that societies entertain about their biology. Nevertheless she goes on to claim that in order to change this situation we must change both social ideas and material reality, both the 'cultural view' of women and 'social actuality' (Ortner 1974:87). It has been suggested (Friedl 1975) that Ortner envisages such changes resulting from a more equal participation by men in childcare. But it is hard to see how such a change could take place given Ortner's account of human sex roles since she implies that women's role in childcare is eternally given by the social construction of women's biology:

> 'Since the mother's body goes through its lactation processes in direct relation to a pregnancy with a particular child, the relationship of nursing between mother and child is seen as a natural bond . . . Mothers and their children, according to cultural reasoning, belong together . . . Mother is the obvious person for this task of caring for 'children beyond infancy', as an extension of her natural nursing bond with children.' (Ortner 1974:77)

Moreover, even if men did participate more equally in childcare, this would not, according to Ortner's theory, result in changes in the 'cultural view' of women, for she says that this view is given by women's biology, not by their role in childcare which is merely an effect of it.

How might 'genuine change' occur in women's social status given Ortner's version of social constructionism? Presumably such change could occur only if either (1) societies gave up construing women's biology as placing them closer to nature, or (2) societies gave up valuing culture above nature. The first possibility is, however, ruled out because Ortner implies that the facts of women's biology necessarily mean that women will be construed as closer to nature than men. This leaves us with the second possibility. Yet although Ortner implies that the social valuation of culture above nature could be abandoned, since this valuation is merely a product of culture, she also implies that this evaluatory distinction is necessarily and essentially given by the very structure of society itself. She implies that the distinction between nature and culture is a necessary feature of all societies and that only by distinguishing themselves as different from, and above nature do humans constitute themselves as social and cultural beings. As she sees it, therefore, the subordination of women will only end given the overthrow of an evaluatory distinction which is definitive of social and cultural life, not only now but at all times. She thus ends up by trading in biological determinism for another essentially deterministic account of women's destiny. This destiny is, in her view, given by structures of social classification that turn out to be just as inflexible as any of the biological factors adduced by biological determinism to explain women's social status. As has been pointed out in criticism of the Levi Straussian structuralism on which Ortner's theory is based:

> 'We might choose to see all human society structured in contrastive categories of female-male, nature-culture, passive-active, uncontrolled-controlled, and copier-creator. If we do, however, then there logically can be no place for women, except at the most menial levels of production and reproduction, in the development plans of nations.' (MacCormack 1977:99–100)

The social construction of menstruation

It could, perhaps, be retorted at this point that whether we like it or not, women's biological processes generally are construed as rendering them closer to nature than men, that this has indeed been the cause of their social subordination. It has been argued, for instance, not only by Ortner but also by many other writers, that the process of menstruation seems to place women closer to nature than men and that it is this social attitude that has been a major source of women's social subordination. I would now like to consider this example of social constructionism; that is, the view that sexual divisions in society are shaped by the way societies construe the biological process of menstruation.

Certainly menstrual blood has been, and still is apparently sometimes construed as placing women closer to nature, at least in the sense of representing a threat to culture and to the attempt by culture to bring nature under its 'control' (Silverman 1977 : 93). Pliny, for instance, described menstrual blood as a 'fatal poison, corrupting and decomposing urine, depriving seeds of their fecundity, destroying insects, blasting garden flowers and grasses, causing fruits to fall from branches, dulling razors' (Weideger 1975 : 85). Pliny's opinion on this matter has, moreover, been constantly reiterated since he first advanced it; it was repeated, for instance, in medieval texts on women's sexuality (Lemay 1978), and even much more recently the belief that menstruation could damage organic matter – that it could turn meat bad – was used as a reason why menstruating women should not salt meat (Dalton 1969). Even today, such beliefs are still given credence in popular astrological books. One such book, for instance, asserts that 'A swarm of bees will die at once if even looked upon by a menstruous woman' (Snow and Johnson 1978 : 70).

Not only is menstrual blood thus regarded by some as dangerous to the cultural control of nature, but such social attitudes towards this aspect of female biology also appear to contribute to women's social subordination, or at least to their exclusion from certain aspects of social life. The exclusion of menstruating women from agricultural activity in certain societies is, for instance, accounted for in terms of the belief that menstruation damages the crops. Similarly, even in relatively recent times, the exclusion of menstruating women from work in wineries and breweries was explained in terms of the possibility that they might harm the wine or beer. In traditional Jewish society, apparently, women were excluded from the temple and indeed from political and economic life on the grounds that menstruation rendered them polluting. And in some societies women are entirely secluded away from society in menstrual huts when they are menstruating lest they wreak havoc both on themselves and on the rest of society.[1]

It could be argued against the social construction account of the influence of menstruation on women's social role that this influence is not mediated by social construction, but is instead the reflection of the way menstruation directly affects behaviour. It has been suggested that menstrual avoidance taboos are determined directly by biology – not by the social construction of biology, but by biologically given, harmful effects of menstruation. Women, it is suggested (De Rios 1978), are excluded in hunting and gathering societies from the prestigious activity of hunting, and from contact with hunting weapons, because the odour of menstrual blood would alert the prey and would therefore ruin the hunt. Others have suggested that menstruation results directly in marked psychological disability ranging from lack of concentration at work to psychosis, violent crime, and suicide! It is because the menstrual cycle directly causes such devastating effects, suggests Dalton, that societies have instituted taboos to ward off its dangers.

But even if menstruation directly caused the dire physical and mental effects claimed by these authors – and the evidence certainly does not seem very convincing (Hollingworth 1914; Parlee 1973) this could not explain the complexity of menstrual avoidance taboos (Montgomery 1974). Most of the research on these taboos has therefore been guided by the theory that the menstrual avoidance taboos that circumscribe women's social activity are the effect, not of the direct action of this aspect of female biology on behaviour, but of the psychological or social construction of that biology.

The theory that it is the psychological construction of menstruation that mediates the impact of this aspect of women's biology on their social status usually takes Freud's psychoanalytic theory as a starting point. Freud himself suggested that menstrual taboos are the effect of the way menstrual blood is psychologically interpreted and construed in terms of sadism:

> 'Primitive people cannot dissociate the puzzling phenomenon of this monthly flow of blood from sadistic ideas. Menstruation, especially its first appearance, is interpreted as the bite of some spirit-animal, perhaps as a sign of sexual intercourse with this spirit. Occasionally some report gives grounds for recognizing the spirit as that of an ancestor and then, supported by other findings, we understand that the menstruating girl is taboo because she is the property of this ancestral spirit.' (Freud 1918: 197)

Freud was not, however, entirely happy with this explanation since, he said, 'horror of blood' could not explain the social practices of circumcision or clitoridectomy. However, he did claim it as a general principle that: 'Wherever primitive man has set up a taboo he

explain, as Ortner claims, the origins of sexual divisions in society in the first place. It is not the social construction of menstruation, for instance, that explains variations in the degree to which this construction is emphasized, or therefore the degree to which women are excluded on the basis of menstrual taboos from social life. Rather it is existing sexual inequalities that determine the degree to which menstrual taboos – the degree to which the social construction of menstruation as polluting – is emphasized.

The cause of sexual inequality is not to be found in the *ideas* that societies entertain about women's biology. It is instead to be found in the *material realities* of social existence. I recognize that this is a contentious point of view. However, it would take us too far away from the matter in hand to defend this thesis in detail here. I have, however, tried to show that, at least as far as ideas about menstruation are concerned, these ideas cannot adequately explain the origins of sexual divisions in society. It therefore seems more plausible to suggest, as Zelman does, that these divisions in non-industrial societies originate and vary according to the following kinds of material factor: whether the environment favours subsistence primarily through hunting or horticulture, whether there is local competition for resources, and whether the society is engaged in warfare.

The social construction of menstruation as reinforcing sexual inequality

If we turn now to an examination of the way social attitudes and prejudices about menstruation have been variously emphasized or played down in our own industrial society, we find that even if these attitudes do not explain the origin and character of sexual divisions in society, they do not merely symbolize them either. Such attitudes are also, as we shall see, used to maintain and reinforce those divisions, particularly when they appear to be threatened. Douglas points out that menstrual taboos are used in this way in non-industrial societies, that they tend to be exaggerated where existing relations between the sexes are perceived as contradictory, and as under threat.

Similar conditions in our own society have repeatedly led to the exaggeration of social prejudices about menstruation. Faced with the threat that the women's movement might disrupt existing relations between the sexes, many writers have appealed to prevalent ideas about the harmful effects of menstruation in order to try to reinforce traditional sex roles. Many men of the professional middle class feared, and still fear, that if women are given equal access to professional jobs they will compete with and indeed sometimes oust men from those jobs; that sexual equality therefore threatens men's prerogatives in the professions. In attempting to ward off this threat, and in trying to shore up the exclusion of women from the professions, many conservatives appealed and continue to appeal to prevalent social beliefs about the harmful effects of menstruation to bolster their case for sexual discrimination. In 1873, for instance, the Obstetrical Society of London voted against admitting women to their society on the ground that women were 'unfit to bear the physical fatigues and the mental anxieties of obstetrical practice at menstrual periods, during pregnancy and puerperality' (Tilt 1874: 73). A surgeon of the same era similarly appealed to the contemporary taboo on menstruating women curing meat in order to bolster his argument that women be excluded from the profession of surgery:

> 'If such bad results accrue from a woman curing dead meat whilst she is menstruating, what would result, under similar conditions, from her attempt to cure living flesh in her midwifery or surgical practice?' (Anonymous 1878: 590)

Showalter and Showalter argue that such quotations demonstrate that scientific knowledge 'reflects, rather than determines, the moral biases of an era' (Showalter and Showalter 1970: 88). But in reiterating these biases and taboos these doctors were not merely reflecting folk wisdom. They were seeking to use that wisdom to reinforce and shore up male privilege in the face of threats to it from the women's movement.

The fact that scientists used 'folk wisdom' on the matter of menstruation in a selective fashion also demonstrates that they were not simply reflecting attitudes about menstruation. They used the current social construction of this aspect of women's biology where it served them, not where it did not. A Dr Storer, for instance, appealed to social attitudes about the harmful effects of menstruation to argue against women's entry into the medical profession: 'neither life nor limb,' he said, was safe if submitted to women during their 'periodical infirmity'. On the other hand, he also claimed that appeal should not be made to social attitudes about the harmful effects of menstruation when it came to judging women accused of performing abortions. One can, he said, 'hardly allow to a female physician convicted of criminal abortion the plea that the act was committed during the temporary insanity of her menstruation' (Storer 1868 : 98).

Today, just as in the nineteenth century, scientists appeal to social prejudice about the ill-effects of menstruation when it suits their interests to do so, and at the same time seek to dispel these same prejudices where they do not serve their interests. While, as we shall see later, scientists have often criticized appeals to social beliefs about the ill-effects of menstruation where such appeals are made in the service of securing women certain improvements in their working conditions, they have often been happy to appeal to these self-same beliefs where they have seemed serviceable to the defence of men's privileges at work. In the latter cause, for instance, we find an American doctor appealing in the following terms to social prejudice about the harmful effects of menstruation (and menopause):

> 'If you had an investment in a bank, you wouldn't want the president of your bank making a loan under these raging hormonal influences at that particular moment. Suppose we had a President in the White House, a menopausal woman President, who had to make the decision of the Bay of Pigs, which was, of course, a bad one, or the Russian contretemps with Cuba at that time?' (Paige 1973 : 44)[2]

(Would that all the opponents of feminism would couch their opposition in such obviously self-defeating terms!)

The fact that the opponents of sexual equality have so frequently sought to justify it by appeal to social prejudice about menstruation has stimulated much useful research into the validity of these prejudices. Liberal-minded supporters of the women's movement (e.g. Hollingworth 1914; Parlee 1973) have repeatedly shown, on the basis of such research, that there is little evidence that menstruation does much harm either physically or mentally. They have argued, on this basis, that there is therefore no ground for the wholesale negative social construction of menstruation nor, therefore, for those arguments that defend sexual discrimination by appeal to this construction of it. Radical feminists have also argued that we should seek to overturn the prevalent negative construction of menstruation and thereby fight the sexual discrimination which it is used to support. Unlike liberal feminists, however, who have used scientific research to attempt to demystify menstruation and its effects, some radical feminists have sought to combat the present negative valuation of menstruation simply by reversing this valuation, by giving menstruation a positive social value.

This latter position has been adopted by a number of feminist artists who seek to celebrate menstruation in their art to counter negative social attitudes about it. Judy Chicago, for instance, in her painting 'Menstrual Bathroom', depicts a shelf with Tampax boxes, a wastebasket overflowing with used sanitary towels and a saturated tampon on the floor. Similarly, Isabel Welsh includes in her 'menstrual show' technicolour slides of her using Tampax and ends the show with a 'symbolic tasting of menstrual blood'. Such 'artistic creations', it has been claimed, represent 'a great breakthrough toward ending menstrual taboos' (Delaney, Lupton, and Toth 1976 : 241).

It is certainly important that women be freed from unwarranted negative attitudes towards their bodies. It is doubtful, however, whether the art of Judy Chicago and of Isabel Welsh will contribute much to this cause, since their art obtains its effect largely by flouting menstrual taboos and, in effect, by reinforcing the onlookers' negative attitudes about menstruation. Furthermore, as de Beauvoir says in discussing a similar movement among French writers, although

> 'It's good to demand that a woman should not be made to feel degraded by . . . her monthly period . . . there is no reason at all to fall into some wild narcissism, and build, on the basis of these givens, a system which would be the culture and the life of woman.' (Simons and Benjamin 1979 : 342)

Such an approach to the woman question means, as she points out, 'falling once more into the masculine trap of wishing to enclose ourselves in our differences' (Simons and Benjamin 1979 : 342).

Moreover such celebrations of menstruation do not do away with the fact that menstruation does, in fact, have negative aspects and causes some women real physical and mental distress. Neither the liberal nor the radical feminist critique of the negative social construction of menstruation, and of the way it has been used against women's interests, does justice to the fact that menstruation does have negative side-effects. Both critiques are ultimately derived from abstract principle, rather than from consideration of the actual effects of menstruation. Some liberal feminists approach the problem of menstruation by adopting the principle that women's interests are always best served by denying that any aspect of women's biology need in any way significantly affect women's social situation. On the basis of this principle it has been argued that women do not suffer from menstruation, that there is therefore absolutely no need for employers to take account of this aspect of their biology. This argument, however, is based less on a consideration of the actual effects of menstruation than on the fact that menstruation has been, and continues to be used to attempt to legitimate sexual discrimination. The approach of some radical feminists to the problem of menstruation, on the other hand, relies on the equally abstract principle that women's interests are always best served by insisting on the positive character of all aspects of women's biology, including menstruation. This latter insistence goes along with the claim that women of all nations and of all social classes have essentially identical interests, that this identity is given by their shared biology.

This last point may be illustrated by reference to what one could refer to as the case of the Baltimore Bleed-In. In their book, *The Curse*, Delaney and her co-authors applaud the art of Judy Chicago and of Isabel Welsh described above. They claim that literary creations in the same genre have also served to give menstruation 'positive mythic overtones'. Writers, they say, who celebrate menstruation in their work thereby make it a 'heroic ritual shared by the community of women, connecting them with the rhythm of nature and with each other' (Delaney, Lupton, and Toth 1976 : 168). Inspired by such art work these authors decided to hold a 'Bleed-In' in Baltimore:

> 'We chose Friday, July 13, 1973, for our 'Bleed-In' because Friday, the number thirteen, and the full moon (it shone on us that night) are all ancient female symbols . . . Mary Jane had decorated the bathroom with the signs and symbols of menstruation. Large paper flowers were hanging from the mirror and the door. Stained pads (tomato sauce) were lying at random on the floor . . . Red yarn dangled from the rim of the toilet . . . On the wall was a piece of paper titled "Menstrual Graffiti" . . . we all sat in a circle (a female, womb-like form) . . . We told anecdotes of our first periods . . .' (Delaney, Lupton, and Toth 1976 : 241–42)

And the effect of all this? 'By the end of the evening' the Bleed-In had, say its instigators, 'given a new meaning to our idea of sisterhood', for it had, they say, brought its participants 'together with their sisters in ancient Cambodia, pre-historic New Guinea, contemporary Israel' (Delaney, Lupton, and Toth 1976 : 242).

But surely this claim is absurd? Although the goal of sisterhood is certainly laudable it cannot be achieved in this way. Women of different nations and of different classes have differing and often conflicting interests, differences that cannot simply be wished away by glorifying and wallowing in our shared biology. In the long run sisterhood will be achieved only through a recognition of, and through struggle against those factors which currently divide women.

The radical feminist approach to the woman question outlined by Delaney and her co-authors is similar, in this respect, to that advanced both by biological determinism and by structuralists like Sherri Ortner. Like structuralist social constructionism and like biological determinism, radical feminism in this instance wrongly neglects the changing historical and economic determinants of women's social situation. Instead, it conceptualizes that situation as essentially given for all time by biology.

Social constructionism used against women

I shall end this chapter by using the example of menstruation to show how women of different social classes have different interests even in respect of their shared biology. I shall argue that while social constructionism may serve the interests of middle-class women it has also been used to oppose the interests of working-class women. It has often been useful to middle-class women to show that it is not menstruation *per se*, but the social construction of it, that limits their career possibilities. When addressed to the question of women's work in industry, however, this type of social constructionist argument has more often been used against rather than in favour of women's interests.

Let us first consider social constructionism as it relates to menstruation and the social status of middle-class women. In their case, the thesis that it is not menstruation but certain unwarranted social attitudes towards it which limit their professional opportunities certainly has served their interests. In 1934, for instance, it was claimed (Whitehead 1934) that menstruation made women pilots prone to accidents. On the basis of this claim women were denied the

Pilot's Certificate. In this case it did indeed serve the women's cause to demonstrate that this claim rested on an unwarranted negative social construction of menstruation, and not on any solid empirical evidence: a critique of the evidence adduced in favour of this claim resulted in a reversal of the decision not to grant women the Pilot's Certificate (Klein 1971).

Menstruation might very well not be associated with accident-proneness. It does, however, often cause women physical and psychological discomfort. The assertion that menstruation has no debilitating effects (or, at least, none that cannot be abolished through medication), and that the effects of menstruation are less the effect of biology than of the way societies construe this aspect of female biology, has not, however, always served the interests of women working in industry. Instead this assertion has often been made to serve the interests of industry at the cost of the interests of the women employed by it. In the Second World War, for instance, when the economy demanded the full-time participation of many women in the workforce, the ill-effects of menstruation were denied at their expense. Rather than admit that some women did suffer with their menstrual periods and that they should therefore be given time off work on this account, it was urged that menstrual symptoms were often the effect of suggestion emanating from outdated menstrual taboos. Menstrual symptoms, it was implied, were not usually the direct effect of biology but were instead the effect of an unwarranted social construction of that biology. On the basis of this version of social contructionism it was claimed that women should not be granted dispensations by industry on account of menstrual symptoms. Indeed, argued one Georgene Seward (1944), withdrawal of such dispensations demonstrated that menstrual symptoms were the effect of suggestion, since menstrual absenteeism soon diminished when women were no longer paid for menstrual leave!

Similarly in the 1960s and 1970s when the British and American economies both relied heavily on women's participation in industry, writers argued that any physical or psychological problems caused by menstruation could and should be abolished by medication. These writers, like Seward before them, suggested that the belief that women might necessarily require rest on account of menstrual symptoms depended on an unwarranted social construction of menstruation, not on biological fact. Dr Katharina Dalton in England, for instance, criticizes women who seek dispensations from their employers on account of menstrual symptoms. These women, she says, appeal to unwarranted social beliefs about the physical problems necessarily caused by menstruation in pleading their case for improvements in their working conditions. Some employers, she notes, have given into such pressure and now provide their employees with 'luxurious rest rooms for women' (Dalton 1969:125). Better, she says, that the medical advisors to these industries should seek how to eliminate menstrual problems through hormone therapy, or that industry should assign women 'to less skilled jobs such as packing and stacking during their vulnerable days' (Dalton 1979:106).

Dalton's recommendations regarding the treatment of menstrual disorders are governed primarily by concern for the interests of industry rather than for the interests of women themselves. She measures the cost of these disorders in terms of their cost to industry, not to women: 'The cost to industry of menstrual problems . . . is measured in millions of pounds, liras, kroners or dollars, not in terms of human misery, unhappiness or pain' (Dalton 1979:100). And she prefaces her discussion of the treatment of menstrual disorders by asking how industry, rather than its female employees, might be best served in this matter. 'How,' she asks, 'can industry cope with this unnecessary financial burden caused by menstrual problems?' (Dalton 1969:105) Not, she suggests, through giving women time off work for these problems. Even though she acknowledges that this solution might serve women's interests, she rules it out on the grounds that it would not serve the interests of industry:

> 'The idea of women going into purdah during those [menstrual] days has much to recommend it, but one does wonder what would be the effect on the national productivity if 8 million women employees opted out for a few days each month.' (Dalton 1969:137)

Instead she recommends that industry invest in the development of hormone treatment for its employees, not because this might serve its employees' health interests, but because such treatment would have the effect of speeding up 'the time when such wastage [i.e., 'the cost of menstrual problems to American industry'] will be eliminated' (Dalton 1979:191).

Grace Naismith, in America, advances arguments essentially similar to Dalton's in this context. Like Dalton, she criticizes women who 'consciously or unconsciously use painful menstruation as an excuse for staying at home' (Naismith 1966:174). And, like her, she implies that the belief that for some women menstruation is necessarily painful is based on an unwarranted social construction of their biology. Dysmenorrhea (i.e. painful periods) is, she says, extremely costly to the American economy. Therefore, she argues, employers should have the right to ask women who suffer from dysmenorrhea to get medical treatment including, presumably, the hormonal treatments peddled by Dalton. Again, like Dalton, Naismith's primary concern is with the interests of the employer rather than with those of the employee. In demanding rights for the former she implicitly wants to reduce the rights of the latter, their right to take time off work on account of sickness in the form of dysmenorrhea.

In sum, the social constructionist account of the influence of biology, at least of menstruation, on women's lives does not necessarily serve the interests of women. As advanced by writers like Seward, Dalton, and Naismith this thesis has been aimed at serving the interests of industry rather than those of women. It might be argued that hormonal treatments, in eliminating menstrual disorders, serve the interests *both* of industry *and* of its female employees. But, in fact, there is now considerable evidence to indicate that the kind of hormonal treatment recommended by Dalton has harmful side-effects on women's health (e.g. Seaman and Seaman 1977). If this is the case then such treatments, while serving the interests of industry in speedily returning women with menstrual problems to the workbench, do so at the expense of women's health interests. It might serve women's interests better to give them time off work for menstrual problems rather than subject them to hazardous hormonal treatments. Despite the discovery of this kind of treatment there is still much justice in the precept of the nineteenth-century doctor who argued that for women who suffer at menstruation and who are 'engaged in industrial pursuits or others, under the command of an employer, humanity dictates that rest from work during the period of pain be afforded whenever practicable' (Jacobi 1877:232).

The denial, on grounds of abstract principle, that menstruation has any negative aspects, and the claim that belief in these aspects reflects an unwarranted negative social construction of female biology – a superstitious taboo – does not, then, as is sometimes implied, necessarily serve the interest of women. The social constructionist account of the effect of menstruation on women's social situation sometimes put forward by liberal and radical feminists does not necessarily serve women's interests any more than the structuralist version of social constructionism considered at the beginning of this chapter. Certainly, it is important to dispel unwarranted and superstitious social constructions and beliefs about women's biology. Much of the scientific and often feminist-inspired research into the actual, as opposed to the assumed, effects of menstruation on behaviour has been extremely useful in this way to the cause of women. If this cause is to be fully served, however, it is also essential to acknowledge that biology (menstruation, in this case) does have real effects on women's lives and that these effects are not to be dismissed as merely the result of the ideas that societies entertain about it.

Notes

1 For accounts of these and other examples of menstrual avoidance taboos see Weideger (1975), and Delaney, Lupton, and Toth (1976).
2 These remarks were made in the context of the Democratic Party's 1970 mid-term election campaign (See Fritchey (1970)).

Eight

Freud and feminism[1]

One can understand the motive for claiming, as feminists have, that it is the social construction of biology, not biology itself, that determines women's destiny. Nevertheless, as I argued in the last chapter, biology does have real effects on women's lives. I shall now elaborate this point further by reference to recent feminist reformulations of Freudian theory, reformulations that have been very influential on recent attempts to develop a socialist feminist analysis of women's social role. First, however, I shall try to dispel a common misconception about Freudian theory; namely, that it is a biological determinist account of the influence of biology on women's lives.

Freud and biological determinism

Freud was certainly no great advocate of the feminist cause. He sometimes dismissed 'emancipated' women as motivated by unresolved penis envy (Freud 1918 : 205). He also asserted that, since the incest taboo was not reinforced in girls (as it was in boys) by castration anxiety, girls were therefore 'less ready to submit to the great exigencies of life' (Freud 1925 : 257–58). As a result, he said, 'The work of civilization has become increasingly the business of men, it confronts them with ever more difficult tasks and compels them to carry out instinctual sublimations of which women are little capable' (Freud 1930 : 103). He maintained that 'We must not allow ourselves to be deflected from such conclusions by the denials of the feminists, who are anxious to force us to regard the two sexes as completely equal in position and worth' (Freud 1925:258). Nevertheless, he also implied that these conclusions only applied, in a strict sense, to feminine women and to masculine men, that is, to ideal types – rather than to actual men and women. In actuality, he said,

> 'all human individuals, as a result of their bisexual disposition and of cross-inheritance, combine in themselves both masculine and feminine characteristics, so that pure masculinity and femininity remain theoretical constructions of uncertain content.' (Freud 1925:258)

Darwin had maintained that those behavioural traits that were the effect of natural selection tended to be inherited, more or less equally, by both sexes. Behavioural sex differences, which were the effect of sexual selection, he said, did not appear in the infant. Instead he claimed that they only developed later when the individual reached sexual maturity. It was presumably to this theory of the late appearance of behavioural sex differences in development that Freud wished to appeal when he maintained, 'Psychoanalysis has a common basis with biology, in that it presupposes an original bisexuality in human beings (as in animals)' (Freud 1920b:171). Freud thus situated his theory of sex differences within this Darwinian tradition. He did not believe, as some behavioural scientists (e.g. Hutt 1972) now do, that boys and girls are primed by biological factors operating in utero to be masculine or feminine from birth.

Instead he claimed that boys and girls initially exhibit both masculine and feminine behaviours in infancy. In his view, masculinity and femininity are not mechanistically determined by innate factors in male and female biology respectively. Instead he believed that they develop as a result of the way the child comes to interpret the fact of biological sex difference to itself. It was on the basis of this belief – namely, that behavioural sex differences develop out of the way the child interprets this biological fact – that Freud asserted that, in the matter of female psychology, 'Anatomy is Destiny' (Freud 1924:178).

Many feminists have taken this to mean that Freud subscribed to, and perpetrated a biological determinist account of sex roles (e.g. Freeman 1971; Glennon 1979). Ehrenreich and English, for instance, assert that Freud 'held that the female personality was inherently defective . . . due to the absence of a penis' (Ehrenreich and

English 1973 : 44). But if we look at Freud's actual account of the development of psychological sex differences we find that he did not subscribe to a biologically determinist account of female psychology. Instead he regarded the development of the characteristically female (and male) personality as the effect of the way the child construes her (or his) biology. He did not regard psychological sex differences as the mechanistic effect of biology.[2]

Freud claimed that his psychoanalytic observations, and the direct observation of infants, indicated that boys and girls are psychologically similar in early infancy; that they equally show both active and passive – masculine and feminine – traits throughout the oral and anal phases of development; and that they both adopt their mother as primary love-object during this early period. The evidence indicates, he said, that it is only with the dawning of phallic eroticism in late infancy (i.e. with the beginning of the phallic phase of development) that girls and boys begin to diverge in their psychological development. As the clitoris and penis come to mature as erogenous zones, and become the principal bodily sources of sexual pleasure for the child, so according to Freud, the physical differences between these genitals come to acquire an important significance for the child in shaping its subsequent development. It is the psychological significance which the child attaches to this biological difference, not the direct and mechanistic action of biological factors on psychology which, in Freud's view, determines the subsequent development of femininity and masculinity in the child.

Freud implies that the first discovery by the child of the genital sex difference is always accompanied by preference for the penis as compared to the clitoris. The evidence indicates, he says, that initially boys respond to this discovery by supposing that the girl has a penis which is 'still quite small. But when she gets bigger it'll grow all right' (Freud 1908b : 216), and that the girl similarly supposes that 'later on, when she grows older, she will acquire just as big an appendage as the boy's' (Freud 1924 : 178). As children come to recognize that girls indeed do not have penises they suppose, says Freud, that this is the result of castration; boys assume that the girl had a penis initially but has since been castrated (e.g. Freud 1923 : 144), while the girl supposes that she has been castrated and that her lack of a penis is the result of 'a punishment personal to herself' (Freud 1925 : 253). Freud said of this 'castration complex' that its significance could '*only be rightly appreciated if its origin in the phase of phallic primacy is also taken into account*' (Freud 1923 : 144. Freud's emphasis). It is the conjunction of the castration complex with the phallic phase of development which, in his view, marks the beginning of the psychological differences between the sexes.

Phallic eroticism leads the boy to entertain phallic, rather than oral or anal, desires in relation to his primary love-object, his mother. This brings him into direct rivalry with his father: that is, it launches him into the Oedipus complex. His belief that girls have been castrated, together with the fact that he has himself already been threatened, either explicitly or implicitly, with castration, leads the boy to fear that his father will retaliate against his phallic desire for his mother by castrating him (Freud 1916–17; 1934). He therefore abandons his Oedipus complex and replaces his attachment to his mother by identification with his father. The boy thus comes to embark on his specifically masculine destiny.

For the girl, the dawning of phallic eroticism leads, in Freud's view, to envy of the boy's penis, which because of its greater size appears to her to be superior to the clitoris as an organ of sexual pleasure. Freud maintained that 'penis envy' finally leads to the development in the girl of the 'normal female attitude' (Freud 1931 : 230), to the girl's adoption of her father as primary love-object, and to her developing the motive for maternity. The girl discovers at some point that her lack of a penis is not specific to herself but is universal in women, that she has not been castrated but was born without a penis. This discovery, involving as it does the recognition that her mother also lacks a penis, leads her to change her attitude to her mother: the mother now 'suffers a great depreciation' in her daughter's eyes (Freud 1931 : 233). Freud also claimed that girls regularly hold their mothers responsible for sending them into the world 'so insufficiently equipped' (Freud 1925 : 254). It is these developments, he said, which lead the girl to abandon her mother as primary love-object. She now loosens her attachment to her mother and turns to her father, and in Freud's words, 'gives up her wish for a penis and puts in place of it a wish for a child: and *with that purpose in view* she takes her father as a love-object' (Freud 1925 : 256. Freud's emphasis). In sum, penis envy and its resolution are responsible, in Freud's view, for launching the girl on the path towards her feminine destiny.

Since, for Freud, the development of femininity in the girl is not mechanistically determined by biology but results from the way she responds to, and interprets her biology to herself, so variations in female psychology are explained by him as due to differences between girls in the way they respond to, and interpret, the genital sex difference. The girl may be 'frightened by the comparison of herself with boys' and therefore give up 'her sexuality in general' (Freud 1931 : 229). Alternatively, she may continue to entertain 'the hope of getting a penis' and develop a 'masculinity complex' (Freud 1931 : 229–30). Lastly, as we have seen, she may acknowledge that she lacks a penis, hold her mother responsible for this lack, and thence develop an attachment to her father. Because in Freud's (1933) view girls vary for constitutional and other reasons in the way they construe this biological sex difference, so they vary in the degree to which they become primarily passive and feminine, or primarily active and masculine, in their subsequent development. That is, for Freud, psychological femininity is not a necessary consequence of female biology but is rather one possible and very probable consequence of the way girls interpret their biology to themselves, and deal with the discovery of genital sex difference.

Freud and biological essentialism

Before considering various recent social constructionist reformulations of the above theory I wish to point out that Freud explicitly distinguished his account of the influence of biology on masculinity and femininity from biological essentialism, as from biological determinism. By biological essentialism I mean that theory which holds that biology has endowed women with an essential feminine nature (and men with an essential, and different, masculine nature). This theory was propounded by many of Freud's fellow-psychoanalysts in their attempt to circumvent the phallocentrism of his account of female psychology.

Apropos his theories of the origins of psychological sex differences in the castration complex Freud wrote: 'It is to be anticipated that men analysts with feminist views, as well as our women analysts, will disagree with what I have said' (Freud 1931 : 230, n. 1). Freud's anticipations in this matter proved to be correct. Within the year Ernest Jones had rejected Freud's account of female psychological development in favour of biological essentialism. Summarizing recent writing on the subject, Jones said:

'Two distinct views appear to be held in respect of female sexual

> development, and to bring out the contrast between them I will exaggerate them in the following simple statement. According to one, the girl's sexuality is essentially male to start with (at least as soon as she is weaned), and she is driven into femaleness by failure of the male attitude (disappointment in the clitoris). According to the other, it is essentially female to start with, and she is – more or less temporarily – driven into a phallic maleness by failure of the female attitude.' (Jones 1932:468)

Jones attributed the former view to Freud, and himself adopted the latter view stating that a girl's 'femininity develops progressively from the promptings of an instinctual constitution' (Jones 1935:495). In defence of this position he referred to the work of Melanie Klein (1924) and of Karen Horney (1926), work that has had an important influence on some recent developments in feminist theory.[3]

All three analysts attempted to avoid Freud's phallocentrism (that is, his view that female psychological development is predicated on a high regard for male anatomy) by arguing that female and male psychology have distinct and independent biological determinants from birth. Horney's account of female psychology can be taken as illustrative of this development within psychoanalytic theory. She rejected Freud's theory on the grounds of its male bias and replaced it with a description of femininity as rooted in women's 'specific biological nature' (Horney 1926:17). She also took issue with Freud's claim that a girl's attachment to her father is to be explained as developing out of penis envy. Horney countered Freud's claim in this respect by insisting that 'we must resist the temptation to interpret in the light of penis envy the manifestations of so elementary a principle of nature as that of the mutual attraction of the sexes' (Horney 1926:17). The motive for this resistance was to dislodge penis envy from any central place in her account of female psychology.

Freud himself acknowledged that the assumption that the girl's attraction to her father is given by nature is appealing. He said that 'It would be a solution of ideal simplicity if we could suppose that from a particular age onwards the elementary influence of the mutual attraction between the sexes makes itself felt and impels the small woman towards men' (Freud 1933:119). But, he claimed, the evidence indicates that girls in the first instance adopt their mother as primary love-object, and that their attachment to their fathers only develops later as the result of psychological, not merely biological developments.

Much more recently the French psychoanalyst, Luce Irigaray, has developed a similar psychoanalytic and feminist critique of Freudian phallocentrism to that put forward earlier by Horney. She maintains that female psychology is not to be explained in terms of a response to male anatomy. Femininity, she says, is essentially constituted by female biology, by the 'two lips of the female sex' (Irigaray 1977 : 64).[4] Furthermore she maintains that women have a 'specific female desire', and that Freud's rejection of the notion of a feminine libido is simply an effect of his patriarchal attitudes. She does not, however, provide any evidence to show that femininity is, in fact, essentially constituted by biology – let alone that it is constituted by 'two lips' – or that there is a feminine libido. What evidence there is contradicts her claim. Freud acknowledges that it is feasible to suppose the existence of distinct masculine and feminine libidos but, he says, the evidence shows such a supposition to be false:

> 'It would not be surprising if it were to turn out that each sexuality had its own special libido appropriated to it, so that one sort of libido would pursue the aims of a masculine sexual life and another sort those of a feminine one. But nothing of the kind is true. There is only one libido, which serves both the masculine and feminine sexual function.' (Freud 1933 : 131)

As we saw above, Freud claimed that the evidence indicates that the libido in infant boys and girls 'serves both masculine and feminine sexual functions'. He argues, furthermore, that male and female psychology and sexuality are rooted in this early bisexuality, rather than in any essential femininity or masculinity. Horney and Irigaray provide no satisfactory evidence to substantiate their quite different viewpoint on this matter.[5]

Not only is the evidence in favour of this essentialist account of female development inadequate, but it is also doubtful whether such an account really serves the interests of women even though its authors claim that it does. Horney and Irigaray suggest that the way forward for women is to assert their biologically given, essential femininity. They are alike, too, in bemoaning the alienation of women from this supposed essential femininity, and in attributing this alienation in part to the position of women in male-dominated society. Horney, for instance, argues that girls defend against their sexual feelings for their fathers by adopting a 'fiction of maleness' (Horney 1926 : 17), and that this defence is 'reinforced by the actual social subordination of women' (Horney 1926 : 19–20). Irigaray argues that it is the language generated by male-dominated society that is responsible for women's alienation from their sexuality. This language, she says, is ill-fitted to the articulation of female sexuality because it assumes 'a unity in the subject' whereas there is always 'a plurality in feminine language' (Irigaray 1977 : 64)[6] – this plurality being a necessary consequence of her equation of the female sex with its 'two lips'! Women are accordingly deprived of their desire, says Irigaray, because they are unable to express it in the language of male-dominated society.

Horney and Irigaray both imply that feminists should fight male-domination in society so as to enable women to express their essential femininity. Irigaray maintains that 'it is in order to establish their differences that women are claiming their rights' (Irigaray 1977 : 68).[7] The slogan 'equal but different', though it certainly rallies support from some within today's women's movement, will not however ultimately serve women. It is, I would suggest, unrealistic to seek for sexual equality on the condition that women occupy an essentially different and separate place in society from that occupied by men. I shall argue this point in detail in the next chapter and will now turn to an examination of the social constructionist critique of Freud.

The liberal feminist critique of Freud

Whereas biological essentialists seek social changes that will enable women to assert their supposedly biologically given essential femininity, other feminists point out that there is no evidence that biology has endowed women with an essentially different character from men. Many liberal and socialist feminists have asserted, on the contrary, that biology only influences female psychology indirectly via the way it is constructed within existing society. Women's position in society can, in their view, be altered towards greater equality with men given appropriate changes in the way biological sex differences are socially constructed.

Feminist writers who adopt this latter view differ, however, as to whether they adopt an 'environmentalist' or a 'structuralist' perspective on the way social factors influence the construction of biological sex differences. Some writers imply that this construction varies as a simple function of variation in environmental conditions in societies. Others, by contrast, deny that such environmental variation necessarily leads to variation in the social construction of biological sex differences. Instead they argue that this construction can only be altered given appropriate changes in the underlying social 'structure', which they imply is something that is not automatically changed by variation in the social environment. Indeed, they argue that the elements of social structure that determine the construction of biological sex differences are often essentially identical in different societies despite considerable variation between these societies in the character of their social environments. The difference between these two accounts of the social construction of biological sex differences is reflected in two rather different feminist responses to Freud's account of female psychology – perspectives which have been associated with liberal and socialist feminist theory respectively.

I shall first consider the liberal feminist critique of Freud. Many of these feminists, particularly in the 1960s, took issue with Freud's claim that female psychology is based on envy of the greater size and visibility of the penis. They argued that in so far as women envy men their penises this is only because they envy the power and status of men symbolized by the penis in male-dominated society, and not because they envy the physical chararacterics of the penis *per se*. These writers implied that as the social environment changed towards one of greater equality between the sexes so penis envy would decrease. Eva Figes, for instance, anticipated that 'in a society where all the material rewards did not go to those endowed with penises there would be no natural envy of that regalia' (Figes 1970: 144). Others claimed that women's social status relative to men had already clearly improved since Freud's day and that accordingly, the motive for penis envy had already diminished. Betty Friedan, for instance, argued that

> 'In the light of our new knowledge of cultural processes and of human growth, one would assume that women who grew up with the rights and freedom and education that Victorian women were denied would be different from the women Freud tried to cure. One would assume that they would have much less reason to envy men.' (Friedan 1965: 106)

Friedan's basic assumption was that environmental change consequent on equal rights legislation and provision would lead automatically and directly to changes in female psychology. Her critique of Freud's account of female psychology is appealing in so far as one believes that women's full emancipation can be achieved simply through instituting social reforms within the existing social structure. It is not, however, so appealing if one believes that reform is a necessary but not a sufficient condition for the achievement of full emancipation.

Furthermore, there are biological as well as ideological reasons for rejecting these liberal feminist accounts of female psychology, for they underestimate the part played by biology, and by sexuality, in shaping psychology. Writers like Figes and Friedan argue that the psychological significance of the penis arises primarily because of its symbolic value in male-dominated society. They entirely neglect to mention that the penis also has psychological significance because of its biologically given erogenous character. Neglect of this factor has led such writers (e.g. Thompson 1943) to suppose that in a society that was dominated by women, rather than by men, breast envy would simply replace penis envy. But in any society, the relative significance given by psychology to the breast and to the penis must be determined not only by the relative social status of men and women in that society, but also, among other things, by the different degrees to which biology has endowed these different parts of the body with erogenous potential. Undoubtedly social factors interact with, and reinforce, the psychological effects of this biologically given, sexual factor. But any account of female psychology that neglects the influence on psychology of this latter biological factor will necessarily be an incomplete account of psychology.

The socialist feminist critique of Freud

The discussions of Freud's work provided by Juliet Mitchell and, more recently, by Ros Coward have proved more appealing than the above, environmentalist accounts of Freud to those who seek to develop a socialist feminist analysis of women's destiny. Mitchell and Coward are committed to the view that psychology is essentially determined by social structure rather than by either biological factors, or variations in environmental conditions *within* a given social structure. It is this theoretical perspective that influences their reading of Freud.

They argue that Freud's account of female psychology is consistent with their version of structuralism – that it describes the construction of female consciousness under a patriarchal social structure – and that it should be accepted on this theoretical ground.

In appropriating Freud for their position within feminism these writers have drawn on Lacan's account of Freud's theory. This account is attractive to these writers because it presents the castration and Oedipus complexes as descriptions of the way the child's consciousness is shaped by social structure. The Oedipus complex has been variously interpreted by Lacan's followers (e.g. Althusser 1971) as marking the entry of the child into the Cultural, or into the Symbolic Order. Coward equates these two Orders and writes of the 'construction of sexual differences in cultural, i.e. symbolic, relations' (Coward, Lipshitz, and Cowie 1976:9). Her account of Freud is, however, primarily in symbolic terms and differs, in this respect, from Mitchell's account which is primarily in cultural terms. Since the emphasis of these two accounts is thus somewhat different I shall deal with them separately.

Mitchell maintains that in describing the castration and Oedipus complexes Freud was describing the way the child comes to find its place in patriarchy. She uses the term 'patriarchy' variously to refer to any society the structure of which is given by the 'law of the father' (Mitchell 1974: xvi), or to refer to any society in which men have the power and in which the kinship system involves the exchange of women by men. Within patriarchy, she says, the Oedipus and castration complexes have the effect of forcing the boy to acknowledge the law of the father – the father's 'function' which is 'to be the phallus for the mother' – and of making him situate himself as heir to this law – of learning that, one day, he 'will accede to the father's function' (Mitchell 1974: 397). For the girl these complexes have the effect, in Mitchell's view, of marking her 'acceptance of her inferior, feminine place in patriarchal society' (Mitchell 1974: 366). In sum, Mitchell maintains that the Oedipus complex as described by Freud correctly

> 'reflects the original exogamous incest taboo, the role of the father, the exchange of women and the consequent difference between the sexes . . . It is specific to nothing but patriarchy which is itself, according to Freud, specific to all human civilization.' (Mitchell 1974:377)

Mitchell agrees with Freud's equation of patriarchy with human society for, she says, recent anthropological evidence indicates that the incest taboo, and the kinship system based on the exchange of women by men, governs marriage within all human societies. Indeed, she claims that all societies must have an incest taboo in order to preserve themselves as integral societies. The taboo forces families to exchange their members and this exchange functions to hold the society together (Levi Strauss 1956). It is for this reason that she maintains that 'the rules of kinship are the society' (Mitchell 1974 : 370), that they constitute 'the elementary form of human society; that distinguishes human society from primate groups' (Mitchell 1974 : 374).

The psychology of women and the way it is given by the castration and Oedipus complexes is, in Mitchell's view, essentially constant within any society which is structurally patriarchal. She claims that 'Differences of class, historical epoch, specific social situations alter the expression of femininity; but in relation to the law of the father, women's position across the board is a comparable one' (Mitchell 1974 : 406). Tinkering with social and environmental conditions within patriarchy, as liberal feminists recommend, might change the 'expression of femininity' but it cannot, in her view, significantly change the position of women. Mitchell therefore rejects the liberal feminist claim that social reforms aimed at giving women a more equal place in society with men could lead to any significant changes in the actual position and psychology of women. Female (and male) psychology within a patriarchal social structure is for her importantly determined by the significance given to the penis within this form of social structure. Variations in degrees of male domination within patriarchal societies do not, in her view, lead to changes in the psychological significance of the penis, or, therefore, to any fundamental change in women's destiny. She accordingly rejects what she refers to as 'certain feminist reductions of any paternal and phallic significance to merely male-dominated cultures' (Mitchell 1974 : 398),[8] and dismisses Betty Friedan's position, which she says implies that female psychology has changed simply 'by virtue of the march of progress' (Mitchell 1974 : 324).

Whereas Friedan implies that social and legal reforms can and have changed the psychology of women, Mitchell's account of female psychology implies that only a revolution in the very structure of society can significantly change women's consciousness. Since this consciousness is, in her view, dictated by patriarchal social structures nothing short of the overthrow of patriarchy itself can change that consciousness. And, since, for Mitchell, patriarchy equals human society, its overthrow must necessarily involve the overthrow of society not only as we know it, but as it has ever been known. One might have thought that this would lead Mitchell to be pessimistic regarding the chances of changing and improving women's social status. In fact, however, she is not in the least pessimistic in this respect. Indeed she argues that the end of patriarchy is at hand because the way the family is constituted within capitalism places an intolerable strain on the incest taboo. 'The proximity and centrality of the tabooed relationship within today's nuclear family,' she says, 'must put a different load on the incest desire' (Mitchell 1974: 377). She claims that the desire for incest is not now sufficiently countered by the kinship law of exogamy, because 'the relationship between two parents and their children assumes a dominant role when the complexity of a class society forces the kinship system to recede' (Mitchell 1974: 378).

For Mitchell the way forward for feminism lies in analyzing the contradiction between 'the capitalist ideology of a supposedly natural nuclear family' and 'the kinship structure as it is articulated in the Oedipus Complex'. She tells us that 'It is, I believe, this contradiction which is already being powerfully felt, that must be analysed and then made use of for the overthrow of patriarchy' (Mitchell 1974: 409). It is, however, entirely unclear from Mitchell's account of psychology, and of the social order, how this experienced contradiction is supposed to bring about the downfall of patriarchy. She not only fails to provide feminists with any leads as to how they might mobilize the actual and experienced contradictions in women's lives towards the overthrow of patriarchy, but she also denies that any environmental changes within patriarchy could significantly alter female psychology and thus implies that political action of any sort is futile. Mitchell thus ends up, in effect, by assimilating Freud's account of female psychology to a version of socialism which can be fairly criticized as essentially utopian.

Mitchell claims to find support in Freud for her view that specific environmental factors within patriarchy do not have a significant effect in changing the character of female psychology. She says that in Freud's account of psychology, 'What actually and specifically happens is nowhere near as important as what is expected to happen in man's general cultural history' (Mitchell 1974 : 64). Certainly Freud found from his psychoanalytic work that the child's phantasies and constructions about reality could have an important effect in shaping its subsequent psychological development. He discovered, for instance, that the phantasy that she had been seduced by her father in infancy could have a very strong influence on a woman's subsequent psychology despite the fact that such seduction had never actually occurred. But Freud also maintained that the phantasy itself had a basis in reality, in the actual and real sexual desires of the girl for her father. Moreover, even though he stressed that the phantasy of seduction could markedly affect the child's subsequent development, Freud did not therefore suppose that whether or not seduction actually occurred was immaterial as regards the child's psychology. Indeed he maintained, 'Where seduction intervenes it invariably disturbs the natural course of the developmental processes, and it often leaves behind extensive and lasting consequences' (Freud 1931 : 232).

Far from dismissing as psychologically insignificant the child's actual experiences, Freud stressed that these experiences could be extremely significant in their psychological effects. He maintained, for instance, that the specific ways in which parents respond to their child's masturbation in the phallic stage can crucially affect its subsequent development. He stressed the importance of 'all the factual details of early masturbation' in determining 'the individual's subsequent neurosis or character' (Freud 1933 : 127). He also regarded the following environmental factors as affecting the speed of the child's psychological development:

> 'the date at which the child's brothers or sisters are born or the time when it discovers the differences between the sexes, or again its direct observations of sexual intercourse or its parents' behaviour in encouraging or repelling it.' (Freud 1931 : 242)

Mitchell fails to mention Freud's views regarding the impact of such environmental factors on psychology. And, in her anti-historicist version of Freud, she also neglects to mention that he regarded the specific historical conditions of his own age as having a very marked effect on female psychology. In fact, in respect of contemporary, 'Victorian' childrearing practices, he maintained, 'The harmful results which the strict demand for [sexual] abstinence before marriage produces in women's natures are quite especially apparent' (Freud 1908a : 197). He maintained that the way in which his society brought up its daughters had quite unnecessarily bad effects on female psychology:

> 'Their upbringing forbids their concerning themselves intellectually with sexual problems . . . and frightens them by condemning such curiosity as unwomanly and a sign of a sinful disposition. In this way they are scared away from *any* form of thinking, and knowledge loses its value for them . . . I do not believe that women's "physiological feeble-mindedness" is to be explained by a biological opposition between intellectual work and sexual activity . . . I think that the undoubted intellectual inferiority of so many women can rather be traced back to the inhibition of thought necessitated by sexual suppression.' (Freud 1908a : 198–99. Freud's emphasis)[9]

Although Freud was ultimately pessimistic about the possibilities of sexual liberation within human civilization he certainly felt there was considerable room for improvement within society as he knew it, and his writing implies some of the ways in which he felt childrearing could be altered so as to improve the psychological lot of women. That is, Freud, unlike Mitchell, certainly envisaged ways in which the lot of women, within what she refers to as patriarchy, could be significantly changed for the better.

Mitchell not only underestimates the extent to which what 'actually and specifically happens' in the child's environment shapes its psychology. She also underestimates the extent to which what happens in the child's physical and biological development shapes its psychology. Again she claims support from Freud here. She maintains, for instance, that 'Freud precisely did *not* believe things were biological, instinctual and changeless: he thought they were cultural' (Mitchell 1974 : 324. Mitchell's emphasis). But Freud clearly did regard 'things' – human psychology at least – as importantly given by biology. The significance given by the child to the genital sex difference was, in his view, importantly affected by a biological factor, namely the physical maturation of the genitals as erogenous zones. Mitchell's underestimation of the place given by Freud to biology also leads her to misrepresent him as providing an 'analytical' rather than a 'developmental' account of psychology. She claims that, 'Above all, Freud's case-histories show that there is little justification for treating the "stages" either as absolutes or as even separate and sequential' (Mitchell 1974 : 27). But Freud did maintain that there were distinct and separable stages in child development and that these stages occurred sequentially, in a sequence timed, in part, by the sequence in the biological maturation of the oral, anal, and then genital regions as erogenous zones. By failing to recognize that Freud was concerned to provide an account of psychology in terms of stages of development Mitchell makes nonsense of his concepts of 'fixation' and 'regression' which are essentially stage concepts.

The details of Freud's account of the influence of these maturational factors on psychological developments may prove ultimately to be incorrect. But any complete account of human psychology must include within it an account of how the biological factor of physical development in childhood is reflected in psychological development. Freud at least seeks to provide some account here of the interaction between biology and psychology. In ignoring this aspect of Freud's account of female psychology, and in rejecting a developmental view of psychology in favour of an analytical account (Mitchell 1974 : 14), Mitchell provides not only a 'partial reading' (Burniston, Mort, and Weedon 1978 : 111) of Freud, but also a necessarily incomplete account of female psychology.

I shall now consider Ros Coward's account of Freud's theory of female psychological development. Like other feminists before her, Coward assesses Freud's theory not so much in terms of its adequacy in dealing with the known facts about female psychology, but in terms of how well it accords with her own theoretical and ideological presuppositions about women's social situation. She dismisses any biological essentialist account of female psychology on purely ideological grounds:

> 'As socialist feminists it would be impossible to accept an idea of a pre-given anatomical identity which was alienated now but would find its true self as soon as capitalism was overthrown, for such an idea denies all possibility of change.' (Coward 1978a : 44).

Instead she argues that what is needed is a theory of 'the construction of the subject in the social process' (Coward, Lipshitz, and Cowie 1976 : 7). And in so far as she regards Freud as providing such a theory, she accepts his account of female psychology as valid and claims that 'the development of psychoanalysis has made it possible for us to see how women are constructed in a sexuality dominated by the relations of reproduction' (Coward, Lipshitz, and Cowie 1976 : 8). Since, in Coward's view, sexuality within a patriarchal social structure *is* dominated by the 'relations of reproduction', so, for her, Freud's account of female psychology is in effect an account of how women are constructed in patriarchy.

Coward's account of Freud's theory draws very heavily on Lacan's reformulation of it in terms of Saussurian linguistics and in terms of the construction of the subject in symbolic relations. According to Coward, the 'phallus' – that is, the 'figurative representation of the male organ' (Coward, Lipshitz, and Cowie 1976 : 14) – is 'the central term in the entry into these [symbolic] relations' (Coward, Lipshitz, and Cowie 1976 : 13). In her view, which she attributes to de Saussure, words acquire meaning in so far as they signify difference. The phallus has meaning, in these terms, by virtue of the fact that it signifies the genital sex difference. Its function within patriarchy, she says, is to channel desire according to this difference, to ensure the 'repression of any sexuality which exceeds the limits of the genital phase', and to organize 'women's sexuality to the relations of reproduction' (Coward, Lipshitz, and Cowie 1976 : 15).

Freud's account of the castration complex becomes reinterpreted, in Coward's system, as the entry of the child into the patriarchal Symbolic Order. Patriarchy, she says, requires the child to take account of the genital sex difference and to situate itself in relation to the phallic signifier of difference. And, since, in her view, symbolic relations within patriarchy are established around 'possession or non-possession of the phallus' (Coward, Lipshitz, and Cowie 1976 : 13), so the child can only acquire language by locating itself with respect to this difference. Within Coward's terms, the psychological significance of the castration complex for the boy is that it enables him to represent himself in terms of possession of the phallus, 'to represent what is in fact present – his penis – and this representation is the phallus, the mark of difference' (Coward 1978a : 46). Similarly, the girl's castration complex enables her to represent herself in terms of non-possession of the phallus. In sum, Coward maintains that the psychological importance of the genital sex difference is that it enables the child to situate itself in terms of the phallus, and hence enables it to acquire language, and to find its 'position in culture as a sexed subject' (Coward 1978a : 46).

Coward argues that the child's construction of the genital sex difference is entirely determined by the 'pre-existent linguistic and cultural order' (Coward 1978a : 46), by the privilege given by that order to the phallic signifier of difference. She stresses that 'There is simply no way in which he or she [the infant] can 'freely' choose to interpret their sexuality or culture as individuals' (Coward, Lipshitz, and Cowie 1976 : 9). Her account of Freud stands in strong contrast, in this respect, to that of Simone de Beauvoir who shows that the great contribution of Freud, as far as women are concerned, was to demonstrate that 'It is not nature that defines woman; it is she who defines herself by dealing with nature on her own account in her emotional life' (de Beauvoir 1976 : 38).[10] Althusser, Lacan, and their followers, like Coward, reject such an existentialist position[11] and argue that it is illusory to suppose that individuals can freely choose how to 'define' and construct themselves. In effect, these authors replace biological determinism with social determinism. For Coward it is not nature, nor individual choice, but social structure that 'defines woman'.

The different versions of Freud presented by de Beauvoir and by Coward reflect their quite different views regarding the way forward for women. De Beauvoir implies in her writing that this way lies, in part, in women exercising their individual freedom so as to achieve transcendence.[12] Coward, on the other hand, implies that the notion of individual free will is an illusion, that there is nothing the individual woman can do to change the structuring of her consciousness. In her view, the structuring of consciousness can only alter given certain changes in social structure, and the structuring of women's consciousness can only alter given the overthrow of patriarchy. Within patriarchy, says Coward, this consciousness is necessarily structured around 'phallic lack'. She implies that this situation can only change given the achievement of a non-patriarchal social structure in which the phallus no longer has a privileged place in symbolic relations, and therefore in which female consciousness is no longer structured in terms of this symbol. Coward implies that the way forward for women is to fight for the overthrow of patriarchy, for an organization of symbolic relations in which the phallus is no longer a 'central term in the entry into' them. But her account of psychology also in effect precludes the possibility of such structural changes ever being achieved. Since, in her view, the consciousness of anyone growing up within patriarchy is structured in terms of the phallus, and since she also claims that individuals cannot change the structuring even of their own consciousness, it follows that they cannot act so as to change the general structure of symbolic relations within the society as a whole so that the phallus loses its privilege in those relations. In sum, Coward, like Mitchell, envisages the changes in social structure which could lead to changes in female psychology but rules out any means of achieving such changes. Although Coward allies herself with socialism, her analysis of women's position, like Mitchell's, turns out to be in effect essentially utopian.

We saw above that Coward very clearly differs from de Beauvoir regarding the degree to which women can exercise individual freedom in the way they construct their consciousness, and interpret their biology. She is, however, similar to de Beauvoir in that she too bowdlerizes the sexual element from Freud's theory. These two writers agree in viewing the penis as affecting psychological development primarily by way of its symbolic value. In de Beauvoir's view it is because the penis symbolizes the superior 'Being' of man as 'Subject' that it is envied by girls. It is to this factor that she refers when she says that 'The little girl's covetousness, when it exists, results from a previous evaluation of virility' (de Beauvoir 1976: 41). In de Beauvoir's view it is the existential value symbolized by the penis, not its sexual character, which, if anything, is the primary motive for penis envy. For Coward the psychological significance of the penis derives from the place of the phallus in symbolic relations within patriarchy. The symbolic value of the phallus does not, in her view, derive directly from the biologically given sexual character of the penis. It derives rather from the place given by society to this sexual factor. Where, as in patriarchy, the society gives genital eroticism a central place in the organization of sexuality, then the phallus will, by virtue of this social construction of sexuality, have an important symbolic value. The importance of the phallus in 'symbolic relations' is not in Coward's view guaranteed by the biologically given erogenous potential of the penis; the phallus would lose its symbolic value, according to her account, in a society in which sexuality was not organized around genital eroticism or around the 'relations of reproduction'.

The view that the organization of sexuality is determined by culture, not by biology, might suit Coward's theoretical predilections. She does not, however, provide any evidence to substantiate this viewpoint. Freud, on the other hand, does provide evidence for his view that by adulthood the sexual instincts normally come to be organized 'under the *primacy of the genital zones*' (Freud 1922: 245,

Freud's emphasis). He acknowledged that bodily zones, other than the genitals, may retain or acquire primary erogenous significance in adulthood. However, he maintained that this only occurred in cases of fixation of, or regression to, 'infantile tendencies' (Freud 1905:232, n. 1), as in the perversions or neuroses. His psychoanalytic case material also clearly indicated that for neurotics, at least, such an organization of sexuality is the result of their having become alienated, for psychological reasons, from their biologically given genital sexuality. That is, Freud recognized that events within the individual's life could alter the extent to which the genitals were a central focus of her or his sexuality, but he implied, on the basis of his clinical case material, that sexuality is normally organized around genitality and that this is as much due to biological as to social factors. Coward presents no evidence to show that Freud's view is mistaken in this respect although she does implicitly reject his view by maintaining that it is culture, not biology, that determines whether sexuality is organized around genitality.

To conclude: in this chapter I have outlined Freud's account of the way biology influences the development of female psychology, and his view that the child's interpretation of biology – specifically of the genital sex difference – plays a significant part in shaping this development. I have argued that this account is to be distinguished from biological determinism. Many of Freud's feminist critics have variously tried to replace his account of female psychology with either biological essentialist or social constructionist accounts of that psychology; they have supposed either that woman's psychology is essentially given by biology, or that biology only influences female psychology indirectly via the way it is socially constructed. I have suggested that these feminist critiques arise out of, and represent, quite divergent and conflicting viewpoints within the women's movement regarding the influence of biology on women's destiny, and regarding the proper goals for feminism. As accounts of female psychology these differing feminist critiques of Freud are unsatisfactory in that they each, in varying ways, underestimate the direct effect either of biological or of environmental factors on female psychological development. It is to Freud's credit that he did give a place to both these factors in his account of female psychology. The facts may well prove him to have been wrong regarding the relative weights to be given to the contribution of biology and society in shaping the psychology of women. However his attempt to describe how consciousness deals with, and is formed by these factors remains important, particularly to feminists concerned to discover how, despite the basically fixed character of female biology, women's consciousness and with it, their destiny, can be changed. It is for this reason that although they have been wary of psychoanalysis, feminists have returned again and again to a consideration of Freud's account of female psychology.

Notes

1 This chapter is a slightly revised and extended version of an article (Sayers 1979) which first appeared in *Women's Studies International Quarterly*. My thanks to Pergamon Press for permission to use this article here.
2 Although academic psychologists have generally been unsympathetic to psychoanalysis, some of them (e.g. Kohlberg and Ullian 1974; Lewis and Weinraub 1979) have also recently suggested that biology affects gender development, not so much mechanistically, as by the way it is cognitively construed by the infant during early childhood.
3 Klein's influence is evident, for example, in Dorothy Dinnerstein's *The Mermaid and the Minotaur*, and Horney's influence is evident in Adrienne Rich's *Of Woman Born*.
4 Despite the essentialism of Irigaray's analysis it has impressed no less a person than Simone de Beauvoir (Jardine 1979) who is not usually sympathetic to essentialist accounts of women's place in society.
5 Studies of children of ambiguous biological sex have been cited as evidence for the innateness of femininity (e.g. Stoller 1976) but reviews of these studies (e.g. Baker 1980) show that they fail to control adequately for the effect of the environment on gender development and that they therefore fail to prove that femininity is innate.
6 See also Irigaray (1980). The American radical feminist, Mary Daly, asserts by contrast that 'We must learn to dis-spell the language of phallocracy, which keeps us under the spell of brokenness' (Daly 1978:4).
7 This goal is not confined to those feminists who, like Horney and Irigaray, use psychoanalysis to argue for a biological essentialist account of gender difference. Other feminists, who use psychoanalysis to argue for a socialization account of these differences – in terms of the social construction of genital sex difference in infancy, or in terms of the infant's earliest relationships with its parents – also come close, on occasion, to 'opening the way to the reactionary position of "equality in difference" and "complementarity" between men and women' (Wilson 1980:40). Indeed the works of some of these writers (e.g. Dinnerstein

1976) have even been used by Christopher Lasch (1981) to argue that higher valuation should be given to masculinity!

8 Although Mitchell draws a distinction between 'patriarchy' and 'male-dominated culture', the distinction is not entirely clear. Indeed, in so far as she uses the former term to refer to any society in which men have the power, her use of this term appears to be quite indistinguishable from her use of the latter term.

9 Freud (1933) also bemoaned the fact that so many girls repressed their active sexual impulses in early childhood. He did not, as some feminists (e.g. Deckard 1975) suggest, regard normal femininity as necessitating the repression of all impulses towards activity nor did he believe that the pursuit of an intellectual career was necessarily a sign of neurosis in women.

10 It should be pointed out that despite the clear influence of Freud on her earlier work, de Beauvoir now claims that he 'understood absolutely nothing about women' (Jardine 1979 : 228).

11 For an extremely clear account of Lacan's disagreement with existentialism see Turkle (1980).

12 This was de Beauvoir's position in 1949. She later criticized this position as too idealist and stated that she would now adopt a more materialist account of women's place in society (de Beauvoir 1965).

Nine

Biology and mothering

The last two chapters have been devoted primarily to a consideration of the feminist theory that biology influences women's destiny indirectly via the way it is socially constructed. I have also mentioned, in passing, that other feminists propose instead that biology does affect women directly, that biology has endowed women with an essential femininity. Women, they say, have become alienated from their essential nature as a result of living in a male-dominated society and it is the task of feminism to enable women to get back in touch with their biologically given essence by, among other things, persuading society to construe and value femininity and female biology equally with masculinity and male biology. I shall refer to this theory as 'biological essentialism'.

In previous chapters I argued that this theory is based on an invalid premise, namely that biology has endowed women with an eternally fixed feminine character, and that it does not, in the long run, serve women's interests. In this chapter I shall further elaborate these points by reference to recent biological essentialist accounts of mothering. I shall then go on to consider an alternative account of mothering put forward by socialization theory. Lastly, I shall argue that this theory, in rightly rejecting biological essentialism, wrongly neglects the biological processes of childbearing. These processes, I suggest, have an important, though not a determining influence on women's experience as mothers. An examination of psychoanalytic, and psychoanalytically based accounts of women's experience of pregnancy and parturition demonstrates, I shall argue, that to acknowledge that these biological processes influence and interact with women's social situation in no way concedes the case to biological essentialism or to biological determinism.

Biological essentialism

Adrienne Rich has recently argued that women will achieve sexual liberation ultimately only as the result of learning to 'think through the body'. Women, she says, need to connect with 'our great mental capacities, hardly used; our highly developed tactile sense; our genius for close observation; our complicated pain-enduring, multi-pleasured physicality' (Rich 1977 : 290). She explicitly differentiates her position within feminism from that of marxism thus: 'The repossession by women of our bodies will bring far more essential change to human society than the seizing of the means of production by workers' (Rich 1977 : 292). She takes issue with those who argue that full equality between the sexes depends on the socialization of childcare. Indeed she explicitly rejects any proposal for 'state controlled childcare' on the very grounds that marxist feminists (e.g. Benston 1969) advance in its favour, namely that it has been, and can be used to 'introduce large numbers of women into the labour force' (Rich 1977 : xvi).

The American sociologist Alice Rossi also now allies herself with this biological essentialist brand of feminism and rejects her former liberal position regarding sexual equality. She now distinguishes herself from liberal, socialist, and marxist feminists – from those who urge 'equal participation of women and men in the workplace' (Rossi 1977 : 25). She argues instead that men and women do not 'have to be or do the same things' in order to achieve equality (Rossi 1977 : 2). Equality, she says, is quite compatible with the traditional (and, in fact, unequal) division of childcare between the sexes – a division which she now claims is ordained by biologically given differences in the responsiveness of men and women to children.

Many argue, as I shall below, that Rossi's argument is basically reactionary. Nevertheless it does correspond with a real trend within contemporary feminism, namely that of biological essentialism. Rossi is now similar to those in the women's movement who reject the goal of trying to achieve sex equality in the work place, who decry the existing values of occupational life as 'masculinist' (e.g. Boulding 1977 : 228), and who assert in its place the value of 'femininity'. That is, Rossi has now abandoned liberalism in favour of essentialism. She now concurs with those feminists who hold 'that women's particular social experience with nurturance and community has a great deal to contribute to wider social change' (Breines, Cerullo, and Stacey 1978 : 47). And, like other biological essentialists, she urges a celebration of women's supposedly biologically given feminine virtues – virtues which she implies have contributed to 'family support, community-building, and institutional innovation in which women have been for so long engaged' (Rossi 1977 : 25).

In 1964, Rossi recommended that sexual equality would eventually be achieved by getting women and men to participate equally in looking after children. Now she maintains that equality is only to be achieved through asserting the value of women's feminine virtues. Women's childcaring ability is, in her view, one such virtue, one that is not, as she previously thought, the product of socialization but is instead the product of factors specific to women's biology. Biology, she says, has rendered women better able than men to care for children. For this reason she claims that women should not seek equality through demanding public day care provision or through trying to get men to help them out with childcare. Rather, she says, women should band together with other women to care for their children, and should seek equality through getting society to recognize their special, biologically given talents for childrearing, and more generally for 'family support'.

A similar argument has frequently been advanced by conservative writers on the woman question. Many nineteenth-century opponents of women's rights assumed that since women look after children they must, on that account alone, have been endowed by biology with an instinct to help them in this task. Spencer, for instance, asserted that given women's and men's 'respective shares in the rearing and protection of offspring', women must have been endowed more than men with that form of the 'parental instinct' that 'responds to infantile helplessness', that 'doubtless' this biologically given 'specialized instinct' conferred on women 'special aptitudes for dealing with infantine life' (Spencer 1873 : 31, 32). Just as Rossi maintains that women's biologically given talent for responding to infants means that women should not seek equality through entry into the workforce, so Spencer argued, as we have seen, that women's 'parental instinct' meant that they should not seek to enter public life. Indeed he claimed that this instinct would be a veritable liability in public life where it would lead to the fostering of social inadequacy.[1] What was needed he said was that women should properly understand the value of maternity:

> 'If women comprehended all that is contained in the domestic sphere, they would ask no other. If they could see everything which is implied in the right education of children, to a full conception of which no man has yet risen, much less any woman, they would seek no higher function.' (Spencer 1898 : 769).

Spencer was not alone in the nineteenth century in assuming that biology had endowed women with a specialized instinct for looking after children (an instinct which, in his view, could also benefit from education in motherhood). Darwin, himself, favoured the notion of inherited instincts and rejected the associationist doctrine that all social feelings are acquired through experience (Young 1970). 'It can hardly be disputed,' he asserted, 'that the social feelings are instinctive or innate in the lower animals; and why should they not be so in man?' (Darwin 1896 : 98, n. 5). He counted the feeling of the mother for her child as one such instinct, and maintained that the 'maternal instincts' lead women to show 'greater tenderness and less selfishness' and to display 'these qualities towards her infants in an eminent degree' (Darwin 1896 : 563).

Darwin did not deal explicitly with the issue of women's rights. His disciple, the naturalist George Romanes, did however address this issue. Like Spencer, and like modern biological essentialists, he argued that biology had endowed the sexes with different psychological characteristics – that equality was to be achieved by giving equal value to these different traits, rather than by giving women equal access with men to the institutions of education and employment. 'There can be no doubt,' he said, 'that the feminine type is fully equal to the masculine, if indeed it be not superior' (Romanes 1887 : 667). The 'maternal instincts' were, he said, one of the 'strongest of all influences' on the development of the feminine type (Romanes 1887 : 663). And he claimed, on the basis of Lamarckian theory, that these instincts had come to be conferred on women by biology, and that they manifested themselves not only in adulthood but also in 'the fondness of little girls for dolls' (Romanes 1887 : 664).

Various suggestions have been, and continue to be made as to how these feminine characters are realized in female biology. In the past some (e.g. Allan 1869) argued that the maternal instincts were located in the cerebral organs of sense, others (e.g. Van de Warker 1875) that they were located in the reproductive organs, specifically in the uterus (e.g. Thomas 1897). Yet others (McCollum, Orent-Keiles, and Day 1939) even went so far as to suggest that they were dependent on the supply of manganese! Nor did such speculation cease with the general decline of instinct theory in recent biological theorizing. Modern writers might no longer countenance the extravagance of William McDougall (1921), instinct theory's greatest champion, and his claims that each type of human social behaviour was the effect of a specialized biological instinct. They do, however, continue to argue that since women look after children biology must have fitted them for this role (e.g. Goldberg 1977; Tiger and Fox 1974). On this ground they continue to search for data which might show that women have indeed been better fitted by biology for the task of childcare. Sexual equality, they say, is not to be sought through insisting on the sexes sharing more equally in childcare but is instead to be found in giving a higher value to women's traditional childcare role and to the biologically given traits on which this role is supposedly based.

Helen Block Lewis (1976), for instance, urges women to value their traditional role in childcare, a role which she says is devalued only because we live in an exploitative society. Society, she says, should give a higher valuation to 'affectionateness' – a trait which, she claims, has been endowed by biology more on women than on men. Her claim in this respect is not, however, warranted by the evidence she cites. Her only evidence is some data suggestive of infant sex differences in affectionateness. She quotes one investigator who found that newborn girls spend more time in reflex 'smiles' than boys. But as Lewis herself points out, this newborn smile is not a social smile. This finding therefore hardly constitutes a secure basis for concluding, as Lewis does, that girls are innately more affectionate and sociable than boys. The only other evidence that Lewis cites in support of this conclusion is a research report that claims to demonstrate that girls show stranger-anxiety earlier than boys, and another report that one-year-old girls stay closer to their mothers in an experimental nursery than do one-year-old boys. However, as Lewis herself acknowledges the evidence regarding sex differences in stranger-anxiety is equivocal. If one looks at all the reported data on sex differences in affectionateness during infancy one finds that there is no consistent tendency for girls to be more affectionate and sociable than boys or vice versa (e.g. Maccoby and Jacklin 1974; Sayers 1980a).

Others have assumed that female affectionateness and nurturance towards infants is the effect of biological processes occurring in pregnancy and parturition. John Bowlby, for instance, argues that the hormonal changes occurring at these times increase the mother's maternal capacity, and that since 'a substitute cannot be exposed to the same hormonal levels as the natural mother', therefore 'a substitute's mothering responses may well be less strong and less consistently elicited than those of a natural mother'. (Bowlby 1971 : 365–66). He does not, however, provide any data on humans to support this conclusion.

Since then, Alice Rossi has brought together a number of these and other biological arguments about sex roles in childcare to support her claim that these differences are directly determined by physiology. In the first place, she says, one would expect 'innate factors' to be important in fostering the mother-child relationship because this relationship is critical to species survival, particularly among humans where the infant is more immature at birth than among any other species, and because women are more involved than men in the reproductive process. Second, she says, the quality of the mother's responses to her newborn infant, and the fact that 'infant crying stimulates the secretion of oxytocin in the mother which triggers uterine contraction and nipple erection preparatory to nursing' suggests that early mother-infant interaction is characterized by 'unlearned' and, by implication, by biologically determined responses (Rossi 1977 : 6). Third, the evidence from studies of fetally androgenized girls suggests to Rossi that prenatal hormonal influences predispose children to learn 'socially defined appropriate gender behavior' (Rossi 1977 : 12). Lastly, she claims that since the same hormone (i.e. oxytocin) stimulates uterine contractions in sexual intercourse, the contractions of childbirth, and nipple erection during nursing and loveplay, women are therefore physiologically more disposed to respond to the child than are men. On the basis of these considerations, and drawing on the same terminology as sociobiology, Rossi concludes that 'there may be a biologically based potential for heightened maternal investment in the child, at least through the first months of life, that exceeds the potential for investment by men in fatherhood' (Rossi 1977 : 24).

Rossi's arguments for the biological determinants of women's traditional role in childcare are not, however, very sound. First, it is not at all clear how a relative lack of responsiveness in men to infants is supposed to serve species survival. Yet this is a necessary corollary of Rossi's claim that species survival has meant that the mother-child, rather than the father-child, relation is innate. Second, in the absence of any data on the interaction of men with newborn infants, the data on mother-infant interaction does not constitute evidence of *differences* between men and women in their responsiveness to infants. Where such data is available (e.g. Parke and O'Leary 1976) it is found that mothers and fathers are equally nurturant to their newborn infants. Third, the evidence from data on fetally androgenized girls is not, as I have already indicated (see Chapter 5 above) a sure basis for concluding that fetal androgens predispose children towards masculine behaviour. Nor, therefore, is it an adequate basis on which to conclude that the absence of androgens in the fetal development of girls predisposes them towards feminine behaviour. Lastly, even if we assume that all women nurse their infants – and Rossi admits that in the early 1960s, at least, only 25 per cent of mothers breast-fed their babies – the fact that oxytocin stimulates nipple erection does not necessarily ensure that the mother will respond in a motherly way to her child.

Rossi's biological essentialist argument is not, therefore, factually sound. It also fails to serve the interests of women. It is at best a misguided feminism. Rossi offers women the goal of achieving equality through pursuing an essentially different role from men, one that is confined to the home while men's primary role is in the public sector. It is possible that in the past, in societies in which production within the home was equal in importance and worth with production outside the home, women might have realistically hoped to enjoy equal status with men despite the fact that they were confined to a different sphere of influence. That is, Rossi's biological essentialist thesis that women can achieve equality with men through pursuing different activities from them might have been a viable goal in the past. The historical development of our own society, however, has entailed the progressive subordination of domestic labour to social production, the whittling away of tasks previously done in the home, and the progressive assimilation of those tasks by the public sector. As a result, production within the home can no longer be equal in importance and worth with production outside the home. It is therefore no longer realistic to propose, as Rossi does, that women can achieve equal status with men while at the same time being confined to domestic labour. Because of the way the economy has developed historically women can now only hope to achieve full equality with men through participating equally with them in social production – in production outside the home. The only alternative would be to try to reverse the course of past historical development. Certainly some propose this solution. They decry modern industrial development and plead that we should return to earlier patterns of social and economic organization. Such a solution, however, has little to recommend it, since it is illusory, and since it neglects the progressive effects of industrial development on the quality of human life.

Many women recognize that as long as they remain confined to the home by childcare responsibilities they cannot hope substantially to improve their situation. And it is found that many women, when asked, now assert that they would like more adequate day nursery provision to enable them to go out to work (Moss 1976). The failure to make adequate provision in this area is largely due to economic factors. The economic development of our society may be such that women can now only hope to achieve full equality with men by participating equally with them in social production. At the same time the British and American economies are currently unable to absorb the labour now being offered by men let alone by women. It is for this reason that these countries do not make adequate day care provision for children. As one writer points out:

> 'Class society would be working against itself if it were to offer women conditions that were ideal, or almost ideal, for reconciling her childbearing and childraising functions with her occupational activities. Since the system of production in market economies is unable to absorb all the potential labor force, relieving women of their traditional functions would mean a substantial increase in the available workforce.' (Saffioti 1978 : 299)

Adequate provision of public childcare would aggravate the problem of unemployment already besetting the British and American economies. Since these economies do not currently have any interest in expanding the available labour force by encouraging women to enter the labour market they have no interest in providing public day care. Instead it suits their interests better to have women stay out of the labour market to look after the children. In so far as biological essentialism condones and, indeed, supports this state of affairs in terms of women's supposed biological predisposition towards childcare it can hardly be said to serve the interests of women.

Socialization theory

Liberal, socialist, and marxist feminists have rightly rejected biological essentialism on the ground that the evidence simply does not warrant its conclusion that sex differences in 'motherliness' are biologically determined, and on the ground that sexual equality depends on women achieving equal employment with men. They assert instead that sex differences in nurturance are primarily the product of socialization and that it is pure wishful thinking to hope that sexual equality can be achieved through the celebration of women's supposedly biologically given, feminine talents for childcare. Sexual equality is not, they say, to be achieved on the basis of asserting the equal value of women's and men's supposedly essentially different characters. These feminists recognize, as biological essentialists do not, that women's traditional role in childcare constitutes one very real source of women's oppression.

So obvious is this fact that it has been acknowledged by conservative anti-feminists and by feminists (apart from biological essentialists) alike. Tiger and Fox, for instance, assert that women's childcare role is a primary cause of their subordination; that this role has entailed 'the male control of females for sex and dominance, and the female use of the male for impregnation and protection' (Tiger and Fox 1974: 110). Similarly, Robert Ardrey writes:

> 'We have seen that in territorial lemur groups . . . a female may even be the leader. But those were the days before melancholy fortune burdened the primate with children who take forever to grow up; the evolutionary advance may have been of intellectual advantage to primate potentiality as a whole, but it reduced the primate mother to the status of a second-rate citizen.' (Ardrey 1971: 223–24)

Nor is this a new position within the discipline of anthropology which spawned these three writers. It was a position that was constantly reiterated in the early days of the discipline. One contributor to the early volumes of *American Anthropologist* maintained, for instance, that

> 'the very assertion of those bio-psychic individualities in primitive society – such as the contraction of marriage by the male outside of his own group, by force or otherwise, or such as the hardships, under unfavourable local conditions, of providing shelter and nourishment for the young by the female, demanding, as the price of sexual favours, help and protection – have led woman slowly out of bondage of economic care for her family group, but led her into marital bondage.' (Solotaroff 1898 : 241–42)

Although most feminists agree with these anthropologists that women's traditional role in childcare now contributes to women's economic dependence on, and subordination to men, they do not agree with their claim that this dependence and subordination is universally and eternally dictated by the biological needs of children for childcare. In the first place, they point out that such a claim rests on the invalid projection onto pre-class societies of the pattern of organization that obtains in modern industrial societies where women's childcare responsibilities often do make them economically dependent. Leacock demonstrates that in pre-class societies 'the economy did not involve the dependence of the wife and children on the husband. All major food supplies, large game and produce from the fields, were shared among a group of families' (Leacock 1972 : 33). It was only with the development of a mode of production in which goods were produced for exchange that a distinction grew up between a public world of men's work and a private world of women's work. It was only then that women's childcare responsibilities came to render many of them dependent economically on men and on their work outside the home.

Second, if we consider pre-industrial societies we find that although childbearing, lactation, and childcare often limit the economic activities of women (Brown 1971; Rosenblatt and Cunningham 1976), these activities can be, and often are arranged so as to allow women to participate in the economic activity of these societies (Friedl 1975). Similarly, in our own society, the demands of childcare do not necessitate women's economic dependence on, or social subordination to men provided that appropriate alternative arrangements are made for the care of children.

Some years ago Rossi (1964) herself recognized that women's traditional role in childcare did contribute to their subordinate social status. She argued then, as many feminists argue today, that women would only ultimately achieve social equality with men if they could achieve equality of employment with them. This in turn meant, she said, that women would have to be relieved of their traditionally exclusive responsibility for childcare, that nursery care would have to be provided, and that men would have to become equally responsible with women for looking after children.[2] In those days she implicitly rejected the biological essentialist thesis that the unequal division of childcare between the sexes had been ordained for all time by biologically-given sex differences in responsiveness to children. She insisted, instead, that these sex differences were the product of socialization and could be reduced by appropriate changes in the upbringing of boys and girls. Boys, she said, should be trained like girls both at home and at school for childcare activities, so that as men they would be as capable as women of looking after children and of sharing childcare with them.

Rossi was not, of course, alone in advancing this argument; as a result of the widely held theory that traditional sex roles in childcare were the effect of sex-role socialization, much useful research was done, and continues to be done in order to uncover those aspects of parental behaviour, schooling, and the media that contribute to the perpetuation of sex inequality in adult life.[3] As a result, a number of very worthwhile changes have been introduced into contemporary childrearing practices; many parents, for instance, now seek to treat their sons and daughters more equally, many schools are now encouraging both boys and girls to pursue activities which were previously sex-segregated, and many children's stories and films consciously strive to avoid perpetuating traditional stereotypes regarding women's childcare role.

Although some research (e.g. Flerx, Fidler, and Rogers 1976) has shown that such changes in sex-role socialization may produce short-term alteration in sex-role attitudes, it is not clear that they can, on their own, undo the differences between men and women in their attitudes and their actual behaviour in regard to childcare. Nor is it clear exactly how the images of childcare conveyed to girls through traditional sex-role socialization 'coerces' (Peck and Senderowitz 1974) them into acquiring the childcare skills purveyed by that socialization. It has been suggested (Mischel 1970) that children simply learn these skills through observing and copying the childcare behaviours that they have seen rewarded for their own sex. Children do not, however, exactly mimic the sex-role behaviours they observe around them. It is clear that their own sex-role behaviour is also affected by the way they actively interpret, rather than passively reflect, the behaviour of those around them (Maccoby and Jacklin 1974).

Nancy Chodorow has recently criticized the above accounts of the transmission of traditional childcaring roles across generations. She points out that childcare is not, as sex-role socialization theory implies, 'simply a set of behaviors'. Rather, she says, it involves 'participation in an interpersonal, diffuse, affective relationship', one that cannot simply be acquired through reinforcement, through enjoining girls to act in a motherly way:

> 'Whether or not men in particular or society at large – through media, income distribution, welfare policies, and schools – enforce women's mothering, and expect or require a woman to care for her child, they cannot require or force her to provide adequate parenting unless she, to some degree and on some unconscious or conscious level, has the capacity and sense of self as maternal to do so.' (Chodorow 1978 : 33)

Rejecting social learning theory, Chodorow turns to psychoanalytic object relations theory and uses its account of child development as a framework for understanding and explaining the roots of the existing division of childcare between the sexes. She assumes, without question, that this division is based on a real difference between men and women in their motivation and capacity for looking after children and then goes on to look for possible causes of this difference. She argues that it results from the fact that infants are primarily cared for by women which, she says, leads to girls having a greater sense of identification and mergence with their earliest caretaker (because she is of the same sex) than do boys. This early psychological sex difference persists, she claims, through the oedipal and adolescent stages of development, such that girls arrive at adulthood still needing to recreate their early experience of mergence and identification – a need that men, as a result of their childrearing, are ill-equipped to fulfil. In consequence, says Chodorow, women have to seek elsewhere for satisfaction of this need and they find it in having children of their own. She concludes that women's deep-rooted sense of relatedness not only creates a desire in them for children but also makes them more able and willing than men to look after the children once they are born.

As well as criticizing the sex-role socialization theory of liberal feminism for not adequately explaining how sex differences in mothering capacity are acquired during childhood, Chodorow also criticizes this theory for failing to explain why the sexes have been socialized differently in the first place. Although liberal feminists have done much to expose the multifarious ways in which the sexes are differently socialized in our society, they have often not been concerned to explain the origins of these differences themselves. Socialist feminists, like Chodorow, attempt to go further than liberal feminists in this matter. As well as arguing that differential sex-role socialization is the effect of female-dominated childcare, Chodorow argues that this in turn reflects economic factors:

> 'Family organization, child-care and child-rearing practices, and the relations between women's child care and other responsibilities change in response particularly to changes in the organization of production . . . Sexual inequality is itself embedded in and perpetuated by the organization of these institutions, and is not reproduced according to or solely because of the will of individual actors.' (Chodorow 1978 : 32, 34).

She thus starts her book by suggesting that sex roles in childcare are the effect of the organization of production, that they can ultimately only be changed by changing the organization of production, not by individual volition. Nevertheless she ends her book by suggesting that these roles are to be changed through individual volition, through persuading men to participate more equally with women in looking after children. And, despite her introductory remarks in which she links sexual inequality in childcare with the character of economic production, she concludes by locating the origins of this inequality in mental and psychological phenomena:

> 'The sexual division of labor and women's responsibility for child care are linked to and generate male dominance. Psychologists have demonstrated unequivocally that the very fact of being mothered by a woman generates in men conflicts over masculinity, and a psychology of male dominance, and a need to be superior to women.' (Chodorow 1978 : 214)

In this, Chodorow is similar to Dinnerstein (1976) who also uses psychoanalysis (specifically the insights of Melanie Klein regarding infantile ambivalence) to explain women's oppression in an essentially ahistorical fashion. She attributes this oppression to the universal predominance of women in childcare, to the fact that most people are cared for in infancy primarily by women, and to the possibility that both sexes resolve their infantile feelings of ambivalence towards their first (female) caretaker by rejecting female authority. She suggests that it is for this reason that both women and men look to men for authority and government.

Whereas Chodorow and Dinnerstein suggest that equality in childcare is to be achieved, ultimately, by an act of will, by simply urging men to participate more equally with women in looking after children, marxist feminists have suggested that just as certain ways of organizing production lead to sexual inequality in the first place (as Chodorow acknowledges), so further changes in the organization of production are needed if full equality between the sexes is to be achieved. Margaret Benston, for instance, anticipates that the socialization of all production, including the full socialization of childcare, would lead to a more equal participation of men and women in work and in childcare, that: 'Once women are freed from private production in the home, it will probably be very difficult to maintain for any long period of time a rigid definition of jobs by sex' (Benston 1969 : 22).

Some support for her thesis that sex roles in childcare are related to, and change with changes in the organization of production comes from a study (Barry, Bacon, and Child 1957) that compared socialization practices in a large number of different, non-industrial societies. The findings of this study reveal that the degree to which boys and girls were socialized for different adult roles in childcare was importantly related to the kind of economic activity pursued by the society, and to associated variations in family structure.

Such data casts doubt on the adequacy of the biological essentialist thesis that sex roles in childcare are constant and given for all time by women's supposedly biologically based maternal traits. Given the inadequacy of the data in support of biological essentialism, many feminists, as we have seen, reject this theory and argue instead that sex differences in childcare skills are the effect of sex differences in childhood socialization. Socialist and marxist feminists have gone on to point out that sex-role socialization is, in its turn, the effect of economic factors. Many of these feminists (such as Chodorow and Dinnerstein) also argue that psychoanalytic theory has a contribution to make to our understanding of sex roles in childcare and of

fetus are physically united in pregnancy. Although this fact is constant for all pregnant women, Deutsch claims that women vary as to how they construe this biological constant. She argues that where women are psychologically unwilling to accept their pregnancy they may experience the embryo as a parasite: 'the embryo becomes psychically what it is biologically, an enemy exploiting the maternal organism' (Deutsch 1945:131). But, for Deutsch, this similarity between the actual biological situation of pregnancy and the psychological interpretation of it in terms of parasitism is not mechanically determined by biology but varies according to the woman's psychological attitude towards her pregnancy. For many pregnant women, she says:

> 'the child is psychologically what the fetus is biologically – a part of the mother's own self . . . By tender identification, by perceiving the fruit of her body as part of herself, the pregnant woman is able to transform the "parasite" into a beloved being.' (Deutsch 1945:139)

Again, however, this psychological identification is not, in Deutsch's view, mechanically determined by the biological identity of mother and child in pregnancy. It depends also on other factors in the woman's psychosocial situation: 'the feeling of unity can be achieved only if no disturbing influences assert themselves in the ego' (Deutsch 1945:139).

The difference between the role accorded by the psychoanalyst, Helene Deutsch, and by the biological essentialist, Alice Rossi, to biology in their accounts of the development of the mother-child relation can also be illustrated by comparing their accounts of the relation between lactation and eroticism. Rossi argues that since the hormonal innervation underlying female sexuality and lactation is the same for all women they must therefore all have a greater predisposition towards childcare than men, a predisposition 'rooted in both mammalian physiology and human culture' (Rossi 1977:18). She implies that oxytocin uniformly makes women maternal towards their children, that this hormone results in an 'innate predisposition in the mother to relate intensely to the infant' (Rossi 1977:24).

On the other hand, Deutsch, who also recognizes that there is a biological relation between lactation and eroticism, does not on this account claim that biology automatically makes women more motherly, nor that it makes them automatically able to respond appropriately to their infants in nursing them. Whereas, she says, some women incorporate the sexual sensations accompanying breast feeding 'into the totality of the positive experience, consciously or unconsciously' (Deutsch 1945 : 291), others cannot bear the confusion of conscious sexual emotions with the tender, loving action of nursing' (Deutsch 1945 : 290) and, as a result, react against the child. That is, although both writers claim that the physical connections between breast feeding and sexuality affect the nursing mother and her relation to her infant, they differ in how they view biology as affecting psychology. Rossi implies that this relationship is mechanistic and uniform, whereas Deutsch argues that it is determined by the nature of the woman's psycho-social situation and varies as that situation varies.

Another difference between Deutsch and Rossi relates to the functional significance accorded by them to the biological processes of pregnancy and lactation. Rossi argues that these biological processes automatically adapt women better than men to the function of childcare. Deutsch, as we have seen, recognizes that these biological processes do affect women. She also believes, like Rossi, that women are naturally more motherly than men. She refuses, however, to bolster this belief by appeal to the psycho-biological changes of pregnancy. Motherliness in women is not, in Deutsch's view, the automatic product of female biological processes. Some women, she says, are motherly without ever being pregnant, while others, who have borne children, are not motherly. Although a woman may psychologically identify with her infant during pregnancy, this identification is, she says, necessarily based on fantasy rather than on an object relationship with a real child. In consequence the identification of pregnancy, which is in her view related to, but not determined by the physical unity of mother and fetus in pregnancy, does not constitute the origins of motherliness in the mother even though it has links with the mother's subsequent maternal relations to her child.

In this Deutsch differs from other psychoanalysts (e.g. Benedek 1970a) who suggest that this identification is rooted in the hormonal changes of pregnancy, and that the identification of pregnancy prepares the woman for her motherly function of being empathic with her infant's needs in the newborn period (e.g. Bibring *et al.* 1961; Winnicott 1960). But even these writers acknowledge, as Rossi does not, that the biological processes of pregnancy do not necessarily make women motherly.

Benedek, for instance, argues that the hormonal processes of pregnancy increase the pregnant mother's 'supply of primary narcissism' and that this constitutes a 'well spring of her motherliness' (Benedek 1970a: 141). But, she says, women vary as to whether they enjoy or defend against this narcissistic state. She thus rejects the biological essentialist position which asserts that the hormonal changes of pregnancy automatically render women motherly towards their infants. 'Motherliness in women,' she says, 'is not a simple response to hormonal stimulation brought about by pregnancy and the ensuing necessity to care for the young' (Benedek 1970b: 154).

While asserting the significance of the psycho-biological processes of pregnancy for the mother's subsequent relationship to her child she does not regard this as entailing that women are necessarily more motherly than men. Like women, she says, men draw on their own experiences of being mothered in infancy in being motherly towards their own infants. These experiences are revived in men in so far as they share psychologically in their wives' pregnancy:

> 'The primary drive organisation of the oral phase . . . is the origin of parental tendencies, of motherliness and fatherliness . . . This culturally influenced drive organisation which motivates man's developmental goals toward marriage and fatherhood become integrated with the "regressive" tendencies through identification with his wife during pregnancy. Sharing her fantasies and projecting his own about their yet unborn child, the father revives and relives his identification with his mother and father in their specific developmental significance.' (Benedek 1959: 399)

Lastly, the differences between psychoanalytic and recent biological determinist accounts of the relation between biology and female psychology can also be illustrated by reference to their differing accounts and critiques of the medical management of childbirth. Consider first the biological essentialist account of childbirth. Adrienne Rich, as we saw above, envisages the liberation of women as being achieved through the celebration of female biology and of the essential femininity to which it supposedly gives rise. Like Horney (see Chapter 8 above), she argues that as a result of living in a patriarchal society women have become alienated from their bodies, alienated from their essential femininity. The medical management of childbirth is, she says, one way in which men have sought to divorce women from their bodies, to deprive them of the powers that would otherwise be theirs as the result of their biological role in motherhood. Modern obstetrics, she says, is to be criticized for standing in the way of women 'knowing and coming to terms with their bodies' through childbirth (Rich 1977 : 150).

Similarly, Alice Rossi now regards the proper goal of feminism as that of becoming 'more attuned to the natural environment, in touch with, and respectful of, the rhythm of our own body processes' (Rossi 1977 : 25). And, like Rich, she criticizes modern obstetric practice for interfering with the 'natural' relation of women to their bodies, for interfering with the 'natural process . . . of spontaneous birth' (Rossi 1977 : 20). Anaesthetics, she says, impede the contribution of 'hormonal levels' to the development of mother-child attachment (Rossi 1977 : 19).

Let us now look at the Freudian critique of obstetrics provided by Deutsch. Although she now calls herself a feminist (Cavell 1974), Deutsch was not particularly sympathetic to the feminist cause in 1945 if we are to judge from her writing at that time, in which she constantly equates feminism with pathological, masculine-aggressive tendencies,[4] and in which she repeatedly casts aspersions on the role of the working woman. Like many of today's feminists, however, Deutsch too was critical of modern obstetrics. Unlike Rich and Rossi, however, she did not criticize obstetric practice on the grounds that it alienated women from a biologically given, essential femininity. The idea that women could return to a biologically given essence in childbirth unencumbered by social practice was, in her view, illusory. In all societies, she said, women's experience of their bodies in childbirth is necessarily affected by the social attitudes and practices that inevitably surround childbirth.[5] Women's experience of their bodies, is, in her view, necessarily mediated not only by such social attitudes and practices, but also by each woman's own individual psychology and by the particular social experiences which have shaped that psychology. Deutsch constantly reiterates a psycho-social approach to the understanding of femininity, an approach that rejects the one-sided biological approach of biological essentialism.

Deutsch's objection to current medical practice in relation to childbirth thus starts from a different basis than that of biological essentialism. Whereas Rich criticizes obstetric practice because she rejects any social factor which intrudes between women's biology and their experience of it, Deutsch points out that such factors inevitably do intrude between women's biology and their experience of it. Her objection to modern obstetrics is not to obstetrics *per se* but to the use of anaesthesia when it is not medically necessary. Such a practice, she says, deprives women of the 'feeling of accomplishment' (Deutsch 1945 : 255) resulting from consciously participating in giving birth, a participation that involves dealing actively with eliminative and retentive tendencies, as opposed to the merely passive submission to, and celebration of female biology recommended by essentialist feminists. It is because anaesthesia interferes with this active psychological and physical process that Deutsch criticizes its unnecessary use.

Dana Breen, like Deutsch, has also recently criticized the modern medical management of childbirth, for the way it affects the psychological construction of this biological process, not simply for the fact of its intervening between women and their bodies. She objects to specific details in modern post-natal care which she says are unnecessary and have the effect of reinforcing some women in the belief that their bodies are inadequate. In this context she criticizes feminists, who she says 'have tended to discard the deeper psycho-biologic aspect of a woman's wish to confirm the goodness of her body' (Breen 1978 : 24). In fact, as we have seen, many essentialist feminists both in the past and today have sought to laud women's biology and its 'goodness'. And, as we have also seen, criticism of the medical management of childbirth is not a monopoly of such biological essentialism, of this strand of feminism.

There is some truth in Breen's statement, however, in so far as some feminists opposed to biological essentialism have sometimes tended to 'discard', or at least to neglect, the significance to women of their biological experience in pregnancy. Feminists who espouse a socialization account of motherliness almost uniformly fail to take account of the possibility that women's experience of childbearing plays a significant part in their lives and in their behaviour as mothers. Simone de Beauvoir is virtually alone among feminists opposed to biological essentialism to give weight to *both* the biological *and* the social factors influencing women's experience of motherhood, and it is to her account of mothering that I shall now turn.

De Beauvoir and psychoanalysis

In both emphasizing the importance of biology and rejecting biological essentialism, de Beauvoir draws on the psychoanalytic insight that 'no factor becomes involved in the psychic life without having taken on human significance; it is not the body-object described by biologists that actually exists, but the body as lived in by the subject' (de Beauvoir 1976 : 38).

Although de Beauvoir rejects the psychoanalytic doctrine of the unconscious, and neglects the importance accorded by psychoanalysis to the role of bodily eroticism in psychology (see Chapter 8 above), her account of the roots of motherliness is more akin to the Freudian than to either the biological essentialist or the pure socialization accounts of motherhood outlined above.

Consider, first, the doll play of little girls which is considered by many to be a forerunner of motherliness in women. Many contemporary socialization accounts imply that girls play with dolls more than boys do simply because they are more often given dolls to play with (e.g. Maccoby and Jacklin 1974) and because children's books present doll play as something that only girls indulge in (e.g. Rossi 1964; Sharpe 1976). Other writers in this tradition criticize such accounts. Chodorow, for instance, says 'It is evident that the mothering women do is not something that can be taught simply by giving a girl dolls and telling her that she ought to be mother' (Chodorow 1978 : 33). Chodorow, however, also implies that doll play in girls is based solely on social experience – on the social fact that women predominate in childcare and on the possibility that this fosters greater identification in girls than in boys with the maternal role. Biological determinists (e.g. Hutt 1972; Wilson 1978) on the other hand, suggest that the absence of male hormones in utero may account for doll play in girls, since where such hormones are present prenatally girls do not play so willingly with dolls.

Psychoanalytic accounts of doll play, by contrast, imply that both biological and psychological factors are involved in it. Kestenberg (1956) for instance, argues that, whereas boys discharge sexual desire through masturbation, girls – lacking a comparable 'explorable object' – choose to sublimate erogenous tension through doll play. This choice, she says, is in its turn reinforced by the little girl's identification with the active aspects of her mother's behaviour towards her.

Many psychoanalysts would dispute the details of Kestenberg's account of doll play (details which are not crucial to the immediate discussion). Nevertheless, they would mostly agree with her in providing an account of this behaviour which is distinct from biological determinism. Instead of positing a mechanistic link between biology and doll play, Kestenberg proposes that biology gives rise to erogenous sensations in girls, and that doll play is one, though not an inevitable, way in which girls *choose* to deal with these sensations. Kestenberg both takes account of biology – of the possibility of biologically-given, erotic sensations in infancy, and of the girl's lack of a penis – and at the same time recognizes that these facts do not automatically result in doll play.

De Beauvoir also takes account of the biological fact of genital sex difference, though not of infantile eroticism, in her account of doll play. She suggests that both boys and girls need to be able to project themselves into an object, that boys can project themselves into their penis because it 'can be seen and grasped', while girls, lacking such an organ, use a doll in which to project themselves (de Beauvoir 1976:278).

De Beauvoir thus recognizes that biology might contribute to doll play. However, she goes on to point out, as Kestenberg does not, that the character of this play also has social determinants. In her view, the way the boy incarnates himself in his penis and the way the girl incarnates herself in her doll reflect the fact that the boy has been socialized towards autonomy, the girl towards passivity:

> 'While the boy seeks himself in the penis as an autonomous subject, the little girl coddles her doll and dresses her up as she dreams of being coddled and dressed up herself; inversely, she thinks of herself as a marvellous doll.' (De Beauvoir 1976:279)

Again, although de Beauvoir's account of childbearing departs in several respects from psychoanalytic theory, her account is nevertheless more similar to psychoanalytic, than to either biological essentialist or to socialization accounts of mothering. Biological essentialists and determinists, as we have seen, assert that the hormonal changes accompanying these processes mechanistically fit women for the function of childcare. On the other hand, socialization theory (as outlined above) entirely neglects the biology of child-bearing in its account of sex roles in childcare. De Beauvoir, by contrast, asserts that these biological processes importantly affect women's experience of themselves but at the same time she denies that these processes have an automatic and determining effect on women's psychology or on their role in childcare. Woman's experience of these processes is, she says, mediated by consciousness, and this consciousness is, in turn, affected by women's status in society.

Like Deutsch, de Beauvoir stresses the variation in the ways women experience pregnancy and breastfeeding and, like Deutsch, she adduces this variety as evidence that women's attitude towards mothering is quite different from the maternal instinct in animals. She concludes her discussion of women's experiences in pregnancy and childbirth by saying:

> 'These examples all show that no maternal "instinct" exists: the word hardly applies, in any case, to the human species. The mother's attitude depends on her total situation and her reaction to it. As we have just seen, this is highly variable.' (De Beauvoir 1976:511)

Like Freudian psychoanalysts, de Beauvoir also argues that the physical 'situation' of pregnancy lends itself to certain types of construction – in particular to a narcissistic construction of the self. In common with Deutsch she regards the frequent narcissism of pregnancy as one way women can choose to interpret their biological situation of physically being with child. Deutsch says that some women, who at other times suffer from feelings of inferiority, use this biological experience as a kind of 'vacation from the ego', as a way of gaining a feeling of 'achievement' and of 'importance' (Deutsch 1945:156). She suggests that this interpretation of pregnancy occurs only in relatively few women, in women who suffer from feelings of inferiority. De Beauvoir, on the other hand, suggests that it is a relatively common feeling, that it is engendered both by the biological character of pregnancy, and by the common social situation of women in which they find themselves colluding with men's attempts to make them the 'Other'. The physical nature of pregnancy provides women, she says, with the illusion of having escaped this socially given situation. In pregnancy, claims de Beauvoir, the woman experiences herself as

> 'no longer an object subservient to a subject; she is no longer a subject afflicted with the anxiety that accompanies liberty, she is one with that equivocal reality: life . . . With her ego surrendered,

alienated in her body and in her social dignity, the mother enjoys the comforting illusion of feeling that she is a human being in herself, a value.' (De Beauvoir 1976:496)

In sum, de Beauvoir gives more weight than does Deutsch to the influence of social factors on women's constructions of the biological processes of pregnancy (and she also vehemently disagrees with Deutsch about the issue of anaesthetics in childbirth). Nevertheless, her account of motherhood is much closer to that of Deutsch than it is to either biological essentialist or to current socialization accounts of sex roles in childcare. And this is because she acknowledges, as they do not, that *both* biological *and* social factors contribute to, and affect women's experience of motherhood.

De Beauvoir combines her account of motherhood with the thesis that the way forward for women lies in the achievement of the kind of society that would foster rather than impede their struggle towards self-realization, the goal of 'transcendence', which she puts forward as an ideal for both men and women. In this her feminism is not dissimilar to the liberal ideal put foreward by Friedan (1965) – the ideal of achieving a society in which both women and men can become autonomous individuals. Unlike Friedan, however, she recognizes the importance of biological sex differences. At the same time she does not regard this recognition as a reason for advocating the goal of equality through a celebration of women's supposedly biologically given femininity.

It has been the aim of this chapter to show that sex roles in childcare are related both to social factors and to biological differences between the sexes, differences that influence but do not determine these roles, differences that can be acknowledged without our having to conclude that the goal of equal participation by men and women in public and social life is illusory. I have tried to show that although the biological differences between the sexes in relation to reproduction affect the situation of women and men in society, there is no good evidence to show that these differences determine women's traditional role in childcare.

Notes

1 While writers like Spencer (1873) and Patrick (1895) thus opposed women's suffrage on the grounds that women's maternal instincts would obstruct their political judgment, others (e.g. Brown 1910) opposed it on

the opposite ground, namely that these instincts would be destroyed by the exercise of the vote!

2 Rossi also recommended women to employ other women to look after their children while they went out to work even though she pointed out that childcare was less well paid than industrial work. In effect, then, she recommended women to gain their liberation through exploiting the relatively low-paid childcare services of other women. Similarly, Germaine Greer (1970) envisaged women gaining their liberation by having peasant women look after their children. These examples nicely illustrate the truth of Eleanor Marx's observation, nearly a century ago, that, although bourgeois feminists fail to recognize any class divisions between women, they in fact strive for 'rights that would be an injustice against their working class sisters' (Draper and Lipow 1976 : 220).

3 There is now an enormous literature on these different aspects of childhood socialization. See, for example, Maccoby and Jacklin (1974) and Smith and Lloyd (1978) on the contribution of parents, Parsons *et al.* (1976) and Dweck (1978) on the contribution of teachers, and Weitzman *et al.* (1972), Sternglanz and Serbin (1974), and McGhee and Frueh (1980) on the contribution of the media to sex-role socialization.

4 The characterization of feminists in terms of pathology was not, of course, the invention of psychoanalysis. Nineteenth-century writers also repeatedly characterized feminists in similar terms; as, for instance, 'superficial, flat-chested, thin-voiced Amazons' (Allan 1869 : 212), and as showing 'masculo-femininity (viraginity)' and 'psycho-sexual aberrancy' (Weir 1895 : 819).

5 This same point has been made today by Sally Macintyre (1977).

Ten

Biology and the theories of contemporary feminism

I started this book by showing that many writers, both today and in the past, have berated feminists for not taking account of biology. I have argued that many of these writers have not practiced what they preached in this matter. In claiming to answer the woman question in terms of biology, they have in fact often answered it in terms that depend on appeal to social, not biological, considerations. I then went on to examine the varying ways in which feminists seek to accommodate certain specific aspects of biology within their analysis of women's destiny. In concluding this book I shall now consider how the theories of contemporary feminism – namely, liberalism, marxism, radical feminism, and socialist feminism – have sought to accommodate biology more generally within their accounts of the position of women in society.

First, however, it should be pointed out that the classification of contemporary feminist theories is problematic. There are, for instance, disagreements within each of the different self-described tendencies of feminism, and certain theories – such as those of biological essentialism and of social constructionism (outlined above) – are shared by some within each of these different tendencies. Furthermore, feminists differ as to whether there are any real theoretical differences between them at all. Jo Freeman (1975), for instance, denies that there are any real differences, in this sense, between contemporary feminists. Even when it is acknowledged that there are theoretical and ideological differences within contemporary feminism, there is often disagreement as to how these differences should be classified. This last disagreement itself reflects differences within feminist ideology. Some writers (e.g. Rossi 1969; Morris 1979) assimilate the woman question to the question of racial integration and classify perspectives on the former in terms of liberal perspectives on the latter question. Others (e.g. Chinchilla 1980), who reject liberalism in favour of marxism, accordingly classify feminist theories in marxist rather than in liberal terms. Finally, the labels used to refer to different strands within feminist ideology have changed over time. Thus, for instance, in the late 1960s and early 1970s the term 'feminist', still tarred for many (e.g. Mitchell 1973; Guettel 1974) within the Women's Liberation Movement with its earlier associations with anarchism and essentialism, was then only used to characterize radical groups within the movement. Now, however, this term is used by most writers (e.g. Jaggar and Struhl 1978) to characterize liberal, socialist, and marxist, as well as radical perspectives on the woman question. In examining the various accounts of biology provided by feminist theory, I shall consider separately those accounts provided by what many now agree to term 'liberal', 'classical marxist', 'radical feminist', and 'socialist feminist' theories.

Liberalism

The account of sex roles provided by writers like Betty Friedan (1965) and advanced today by such organizations as the Equal Opportunities Commission in England, and by the National Organization of Women (NOW) in America, derives directly from the doctrines of liberalism first advanced by the bourgeoisie in its struggle to develop a free market society unfettered by feudal ties of birth. Liberals who sympathize with feminism have been quite explicit in maintaining that women should be able to sell their skills on a free market basis equally with men. They oppose sex discrimination on the ground that it hinders the free operation of the labour market. As one nineteenth-century writer put it: since some women are intellectually superior to men they should be 'hindered from the free development and use of their powers solely by reason of their sex'. He argued that 'absolutely free competition between the sexes' should be introduced by removing 'all the bars which at present restrain woman in her industrial life or in her legal, political, and social relations' (Buchner 1893:175).

Liberal feminism and sociobiology both have their roots, historically speaking, in the development of the free market society. I have already tried to show how the sociobiological account of sexual relations relies on the bourgeois ideology of the free market (see Chapter 4 above). Whereas sociobiology elaborates this ideology to develop a biological determinist account of sex roles, liberal feminism has elaborated it to develop a concerted critique of sex discrimination, and of biological arguments designed to justify that discrimination.

The liberal feminist opposition to the claim that biological sex should determine social status is rooted historically in the early opposition of the bourgeoisie to feudalism in which biology, namely the fortunes of birth, determined social status. The bourgeoisie asserted then that social status should be determined by the free market, by the value placed on the individual's labour power relative to that of others in the market, a value that should be determined not by biological inheritance but by the individual's ability and skill. That is, the bourgeoisie asserted that individuals should have the right 'to rise and fall within the marketplace through their own efforts, rather than on the basis of birth' (Zaretsky 1976 : 56). The doctrine of liberalism was, perhaps, given its most forceful expression in the French Revolution. The liberal ideology of the Revolution was reflected in the earliest publications in support of the emancipation of women (Klein 1971 : 94), publications that were directly situated in relation to that event: Mary Wollstonecraft dedicated her book, *A Vindication of the Rights of Women*, to a legislator of the new French regime, and the *Declaration des Droits de Femmes* was written as a submission to the 1798 French Assembly. In her book, Wollstonecraft applied the liberal philosophy of the Revolution – and its tenet that social status should be decided by ability rather than birth – to the woman question, and argued that since women had equal powers of reason (i.e. ability) with men, they should therefore be granted equal rights with them.

Equal rights were, in the course of time, to encompass rights to the vote, to education, and to employment. And just as Wollstonecraft had asserted, in effect, that biological sex did not constitute a proper basis for assigning differential rights in the eighteenth century, so the nineteenth-century advocates of women's rights also asserted that an individual's access to education should not be determined on the basis of biological sex. Advocates of women's rights to higher education argued, for instance, that since women's health was not undermined by education, women's exclusion from it could not legitimately be justified on biological grounds (see Chapter 2 above). Liberal feminists of the twentieth century acknowledge that biology – the fact that women bear children and that children require adult care – is relevant to the issue of equal opportunities in educational and occupational life. Nevertheless they maintain that biology in this sense does not and should not constitute an insuperable obstacle to women's achieving equal rights in these spheres. They point out that these biological facts can be taken care of through the provision of family planning and of adequate public child care. It is for this reason that twentieth-century liberal feminists have often linked their demands for equal rights with demands for free access to birth control (e.g. Eastman 1920), for maternity leave (e.g. Friedan 1965), and for public child care (e.g. Rodman 1915; National Organization of Women 1967).

Given the liberal belief that social status should reflect individual ability it has always been open to the opponents of liberal feminism to challenge it on its own ground by proclaiming that the inequality of the sexes in social status and in educational and occupational achievement does reflect differences in ability, differences determined by biological sex. It was on this basis that in 1762 Rousseau, a champion of liberalism, opposed equal rights for women: 'nature', he said, had rendered women less able than men. Similarly, Herbert Spencer advocated equal rights while opposing women's right to the vote on the ground that biology had made them incapable of exercizing it responsibly. Other self-avowed liberals, both in the nineteenth century and today, likewise asserted and continue to assert that social inequalities between the sexes reflect biologically given sex differences in ability (see Chapters 3, 5, and 6 above).

Liberal feminists share with these biological determinists the premise that social status should reflect individual ability. They have, however, countered biological determinism by attempting to demonstrate that an individual's ability – his or her aggressiveness and competitiveness, or intelligence and spatial skill, for instance – is not determined by biological sex. Any correlation between ability and sex is, they say, primarily the product, not of sex differences in biology, but of differences in education – differences that can be remedied through appropriate changes in childhood socialization. Many of today's liberal feminists, like Wollstonecraft and Mill before them, assert that any inferiority in women's intellects is the effect of their inferior education. Sherman (1977), for instance, counters the argument that women's lack of representation in mathematics-related jobs reflects sex differences in brain organization by arguing that this social inequality is primarily the effect of the fact that girls are systematically socialized away from an interest in mathematics-related activities during childhood. Others, who also adopt the liberal belief that differences in occupational success reflect differences in individual ability, focus not on sex differences in cognitive ability but on sex differences in assertiveness and competitiveness. Such differences, they say, are also not the effect of biology but of differences in sex-role socialization, differences which may be remedied through such measures as assertiveness training (e.g. Osborn and Harris 1975). Yet others argue that even if ability is correlated with sex and is to some extent determined by biological factors associated with sex, this correlation is not perfect. They claim that since there is an overlap in the distribution of men's and women's abilities, biological sex is an inaccurate guide to an individual's ability. And on this basis they claim that access to educational, occupational, or political spheres should not, in all fairness, be determined on the basis of biological sex, but should instead be decided on the basis of individual ability to which biological sex provides an inaccurate guide (e.g. Lincoln 1927; Archer 1976).

Since liberal feminists reject the biological arguments of the liberal anti-feminists as bogus they are led to describe sexual discrimination as unfair and irrational. Given that they reject biological explanations of sexual inequality, however, liberal feminists are unable to explain – at least within the confines of liberal philosophy – why it is that sexual inequality persists in an avowedly liberal society. If one rejects the liberal defence of sexual inequality in terms of biologically given sex differences in ability one cannot explain, in liberal terms, why it is that sexual discrimination continues, despite the fact that writers from the time of Mary Wollstonecraft have pointed out that such discrimination is irrational and inconsistent with the tenets of liberalism. In the same way the rationalists of the eighteenth and nineteenth centuries were unable to explain why the social and political institutions born of the French Revolution, of the supposed 'triumph of reason', should have so singularly failed to embody the liberal ideals on which they were supposedly founded (Engels 1880). In both cases liberalism has been unable to explain why inequality persists after the unreasonableness of that inequality has been clearly pointed out. It is for this reason that, as several commentators (e.g. Chinchilla 1980) have noted, liberal feminism lacks a coherent theory of the causes of sexual inequality. It restricts itself instead to campaigns designed to pressurize existing institutions to be more consistent with their liberal principles, to give legislative force to the principle that all individuals should have equal political and social rights regardless of biological sex. Indeed, so convinced are some liberals that equal rights will be granted as soon as it is shown that they are demanded by the canons of justice and reason, that they have even asserted that 'the easiest target in removing sexual inequality involves legal statute change or judicial interpretation of rights in the public sector' (Rossi 1969 : 81). And this despite the fact that the struggle to gain these rights has always been so hard!

The main recourse for liberal feminists who wish to explain the causes of sexual inequality has been, as we have seen, to argue that this inequality is due to sex differences in ability, and that these differences are in turn due to the different ways in which the two sexes are socialized. Liberalism, however, no more provides a basis for explaining sex differences in socialization than it does for explaining sex discrimination in adult life. Nor does it provide a basis for explaining why it is that liberal anti-feminists have been so vigorous in defending sexual inequality by, for instance, countering environmentalist with biological explanations of that inequality. Marxist theory, on the other hand, does seek to provide an explanation of these phenomena. Marxists agree with liberal feminists that many sex differences in ability are indeed often the effect of sex differences in socialization. However, whereas liberal feminists rest content with this explanation of the genesis of sex differences in ability, marxists have pointed out that sex differences in childhood socialization themselves stand in need of explanation. They suggest (e.g. Griffiths and Saraga 1979) that these latter differences can be explained in terms of the way they serve to reproduce existing sexual divisions in society. They go on to argue that these divisions, in turn, can be accounted for in terms of their relationship to the capitalist mode of production.[1]

Marxist feminists also reject the premise shared by both liberal feminists and liberal anti-feminists: namely, that individual ability – whether expressed in aggressiveness or in such mental traits as spatial ability – determines social status. As Saffioti points out, although this liberal doctrine is itself a product of capitalism it is also negated by it because status in capitalist society is determined primarily by class rather than by individual ability:

> 'For relative surplus value, created by the labor of the actual producer, to be appropriated in the form of profit there must exist a market on which people are able to sell their labor power as free agents. The obverse side of this formal freedom, however, is the division of society into classes and hence the domination of one class by another. The capitalist mode of production with its internal contradictions is a permanent negation of the juridical freedom which it had created to serve its own needs . . . Once the economic foundations of the class division of society had come to the surface, the outlook of individuals tended to be shaped by their position in the class structure.' (Saffioti 1978 : 34)

Marxists have also questioned the assumption of liberal feminism, shared with liberal anti-feminism, namely that a free market in individual talent unfettered by birth necessarily operates for the good of all individuals in society. Writers in the liberal tradition of Wollstonecraft and Mill continue to assert, like them, that the fuller utilization of women's talents will be to the advantage of society in accord with the general thesis of individualism advanced by Adam Smith in 1776 that when each individual pursues his own economic self-interest the 'invisible hand' of Providence working through the market economy will coordinate these selfish strivings for the good of all (Smith 1937 : 423). In fact, as marxists point out, the free market economy of which individualism is an ideological product has not equally benefited all members of society, but has instead resulted in the majority of individuals being exploited by the minority in the capitalist class. And as marxists also point out, the fact of class exploitation means that the granting of equal rights often remains a hollow and formal gesture, since the opportunities to exercise those rights continue to vary systematically with class. Similarly they argue (e.g. Draper and Lipow 1976) that the granting to women of the right, say, to employment or to equal pay does not of itself guarantee women's emancipation, since it does not guarantee that jobs will be available. Nevertheless marxists also maintain that liberal struggles for certain rights – in the past for the right to vote and for the right of assembly, and today for equal pay – are worthwhile because they are a necessary, even if not a sufficient, condition for the achievement of full emancipation.

The fact that liberal feminism accepts the existing institutions of society and simply asks that women have equal access with men to those institutions is sometimes obscured by the fact that liberals (e.g. Rossi 1969; Freeman 1975) often claim (on the basis of the kind of essentialist theory reviewed above, see Chapters 3 and 9) that if feminine and masculine traits were given equal opportunities to shape social institutions this would in effect revolutionize them. But as Marlene Dixon points out:

> 'The claim that status equalization would bring about a "revolution" is of the same order as the claim made by the Suffragists that giving women the vote would usher in an era of world peace. Abolishing discrimination would not lead to a "revolution" in the status of women because it would leave the class structure absolutely untouched.' (Dixon 1978 : 64. See also Gimenez 1975)

The fact that marxist feminists reject the ideological framework of liberal feminism does not, however, mean that they necessarily rule out the possibility of working together with liberal feminists in criticizing the biological arguments of the anti-feminists (even of the anti-feminists within the socialist ranks themselves).[2] Socialists have certainly believed in fighting with liberal feminists for certain reforms within the existing system. Dixon herself maintains that

> 'The discussion of social class and the question of discrimination and legislative reform should not lead one to reject legislative reform. Such reforms as would limit the current abuses of working women are real gains, and worth fighting for.' (Dixon 1972 : 240, n. 6)

Similarly, socialists of an earlier period, although they criticized the stance of the 'women's-rightsers' on many issues, nevertheless supported them in some of their reformist demands, most notably their demand for the suffrage (Draper and Lipow 1976). Socialists have certainly always been wary of joining with bourgeois feminists in pressing for reform – a wariness reflected both in earlier[3] and current marxist discussions of the woman question. On the other hand, as Lenin pointed out in 1902, socialism has also 'always included the struggle for reforms as part of its activities' though 'it subordinates the struggle for reforms, as a part to the whole, to the revolutionary struggle for freedom and socialism' (Lenin 1969 : 62).

Classical marxism

Feminists who have sought to develop a marxist analysis of women's status in society have repeatedly returned to Engels's classic text on this subject. Whereas many of his contemporaries explained the development of family life, and of the social position of men and women in terms drawn from biological evolutionism (see Chapter 3 above), Engels argued that this development was primarily a product not of biology but of historical changes in the mode of production. In this sense, although he offered a materialist account of changes in sex roles, it was a materialism the roots of which were to be found primarily in the material conditions of production rather than in the material conditions of biology. To this extent, E. O. Wilson (1978 : 191) is correct to describe marxism as 'sociobiology without biology', and others are also correct to distinguish marxism from earlier social Darwinist accounts of the development of human society: unlike sociobiology and social Darwinism, marxism does not give priority to biology in its account of society. It is therefore a misrepresentation to characterize marxist writings like *The Origin of the Family* as 'evolutionist', as many do (e.g. Kuhn 1978).

Given the commitment of Marx and Engels to materialism, it would, however, have been surprising had they entirely dismissed biology – one of the most important material factors of human existence – from their account of human society. The fact that, as Delmar says, Engels's achievement was to show 'women's oppression as a problem of history rather than of biology' (Delmar 1976 : 287) has tended to mislead both anti-feminists like E. O. Wilson, and feminists like Firestone, into denying that marxism concerns itself at all with biology. In fact, however, both Marx and Engels were enthusiastic advocates of Darwin's biological ideas. Indeed, Marx dedicated a copy of *Capital* to Darwin (Fay 1980).

It is consistent with his belief in the importance of biology that Engels should have given it a significant place in his account of women's status in society. In the first place, he argued that the fact that women bear children and therefore have a greater degree of certainty than men as to who are their biological offspring had directly affected women's social status. Sociobiologists, as we have seen (Chapter 4 above), assert that the character of human sex roles is essentially fixed for all time by this biological fact. Engels, on the other hand, demonstrates that the way in which this biological fact shapes woman's role in society changes as a function of changes in the mode of production. This role is not, therefore, according to Engels, eternally fixed even though it is directly shaped by a feature of biology which is fixed.

Adopting Morgan's account of 'human progress', Engels argued that in early human history when subsistence depended primarily on the 'appropriation of products in their natural state' (i.e. in hunting and gathering societies), sexual relations were relatively free and were characterized by 'group family' organization (Engels 1972 : 93, 106). In such societies, he said, the biologically given fact that maternity is more certain than paternity led to the development of matrilineal descent systems: 'It is therefore clear that in so far as group marriage prevails, descent can only be proved on the *mother's* side and that only the *female* line is recognized' (Engels 1972 : 106. Engels's emphasis).

Engels goes on to argue that when the mode of production changed the biological fact that women bear children came to have a different consequence for women's social status. The change to an economy based on the domestication and breeding of animals resulted, says Engels, in the production of surplus wealth. The matrilineal descent system bequeathed to this new form of society by earlier forms of society meant, however, that this wealth was the property of the female line: 'When the owner of the herds died, therefore, his herds would go first to his brothers and sisters and to his sister's children, or to the issue of his mother's sisters. But his own children were disinherited' (Engels 1972 : 119). As long as there was no property to inherit, says Engels, matrilineal descent seemed perfectly acceptable. Given an economy that did produce surplus wealth, however, men found this system of descent intolerable. 'As wealth increased,' Engels argues, it 'created an impulse [in men] to exploit this strengthened position in order to overthrow, in favour of his children, the traditional order of inheritance' (Engels 1972 : 119). This resulted, he says, in the overthrow of mother-right and in 'the world historic defeat of the female sex' (Engels 1972 : 120).

The new mode of production also meant that women's biological role in reproduction now had a different effect on their social status. Under the earliest forms of subsistence activity, said Engels, the biological fact that women bore children resulted in matrilineal descent. As the economy came to produce surplus wealth this same biological fact led, in Engels's view, to the subordination of women in marriage. Now that there was wealth to bestow on children, men wished, says Engels, to ensure that this wealth was bestowed on their own children. The fact that biology did not give them complete certainty as to who were their children resulted in men seeking to overcome this biological handicap by instituting social controls over their wives' sexual activity: 'In order to make certain of the wife's fidelity and therefore of the paternity of the children, she is delivered over unconditionally into the power of the husband; if he kills her he is only exercising his rights' (Engels 1972 : 122).

In sum, the subordination of women in sexual relations was, in Engels's view, not a necessary consequence of the biological fact that women bear children (as sociobiologists claim). It resulted instead from the social fact that as the economy came to produce surplus wealth so men wished 'to produce children of undisputed paternity' (Engels 1972 : 125). Since nature makes paternity less certain than maternity, this economic development, says Engels, resulted in men attempting to deal with this biologically given fact by enforcing sexual fidelity in women.

I shall now consider a second way in which Engels believed that the biological fact that women bear children explained their social status. He claimed that this fact explained why it was that the surplus wealth produced by the herding and breeding of animals was the property of men rather than of women, and hence why the accumulation of private property led to the development of patrilineal descent.

Marx and Engels maintained that this physiological sex difference led, from the beginning of human history, to an elementary division of labour between the sexes. Marx, for instance, writes:

> 'Within a family, and after further development within a tribe, there springs up naturally a division of labour, caused by differences of sex and age, a division that is consequently based on a purely physiological foundation.' (Marx 1967 : 351)

This division entailed, said Engels, that from earliest times 'The man fights in the wars, goes hunting and fishing, procures the raw materials of food and the tools necessary for doing so. The woman looks after the house and the preparation of food and clothing, cooks, weaves, sews' (Engels 1972: 218). This results, in its turn, in a division of property between the sexes: 'Each [sex] is owner of the instruments which he or she makes and uses: the man of the weapons, the hunting and fishing implements; the woman of the household gear' (Engels 1972: 218). Since the surplus produced by the domestication of cattle produced wealth in the sphere of men's rather than of women's work this wealth was, says Engels, therefore regarded as the property of men not of women. As a result of this development and of the consequent subordination of women to men the emancipation of women could only be achieved, in Engels's view, by overthrowing the customary sexual division of labour that had until recently come to involve the exclusion of women from social production: 'The emancipation of women will only be possible when woman can take part in production on a large, social scale, and domestic work no longer claims anything but an insignificant amount of her time' (Engels 1972: 221).

This brings us to a third way in which Marx and Engels regarded biology as relevant to women's destiny. We have seen above how Engels used the biological fact that women bear children (and are therefore more certain of their maternity than men of their paternity) together with historical facts about changes in the mode of production to explain the development, first, of matrilineal and, second, of patrilineal descent systems. We have also seen how he used this biological fact to explain why surplus wealth was the property of men, not of women. I now want to consider how the biological fact of sex differences in physical strength entered into the classical marxist account of the ways in which women could be emancipated.

Engels argued that the conditions of modern industry now permitted 'the employment of female labour over a wide range' (Engels 1972: 221). The reason for this was, he said, that technological development had meant that, despite their unequal strength, men and women could now contribute equally to industrial production. Marx described this development in similar terms:

> 'In so far as machinery dispenses with muscular power, it becomes a means of employing labourers of slight muscular strength, and those whose bodily development is incomplete, but whose limbs are all the more supple. The labour of women and children was, therefore, the first thing sought for by capitalists who used machinery.' (Marx 1967: 394)

Similarly, Clara Zetkin was to maintain that modern industrial development had meant that woman had 'become completely equal to the man as labour-power; the machine makes muscular strength unnecessary, and everywhere woman's labour could operate with the same results for production as man's labour' (Zetkin 1896:196).

Zetkin and many of the early socialists were optimistic that because technological development had rendered biological sex differences in strength no longer a crucial factor in the determination of sexual inequality in social production it had thereby rendered women's emancipation a real possibility. They did not, however, claim as some have implied (e.g. Mitchell 1973), that machine technology would on its own result in the liberation of women. Instead, they argued that the employment of women in public industry constituted a premise for women's emancipation (Magas 1971), a starting point for women's struggle by bringing them into the labour force and thus into the working-class struggle (Draper and Lipow 1976).

The position of classical marxism on the woman question has been criticized by many feminists and has given rise to two different ideological tendencies within feminism, namely those of radical and socialist feminism. I shall deal with some of these criticisms later in outlining their perspectives on the bearing of biology on sex roles. I shall restrict myself here to outlining three objections which have been addressed specifically to the place given by Marx and Engels to biology in their account of women's position in society.

First, Engels's accounts of the way the biological fact that women bear children contributed to the development of patrilineage has been criticized on the ground that Engels relied on psychology, rather than on material factors, to explain why it is that men wish 'to ensure that "their" property is handed on to their own children' (Edholm, Harris, and Young 1977:102. See also de Beauvoir 1976). This criticism seems to be just: the explanatory load at this point in Engels's account of the development of monogamy is indeed placed on psychology. He simply asserts that man has 'an *impulse* . . . to overthrow, in favour of his children, the traditional order of inheritance' (Engels 1972:119. My emphasis).

Second, with regard to Marx's and Engels's explanation of the sexual division of labour in terms of biology, it has been pointed out that this division is not solely given by biology. Even in hunting and gathering societies, physiology alone cannot explain all the detailed ways in which the tasks of a society are divided up between the sexes (Levi Strauss 1956).

Third, it is argued against Engels that sex differences in physical strength have never been a major determinant of women's work, or of sexual inequality. Mitchell points out that woman's physical weakness does not remove her from productive work in pre-industrial society, nor has it ever prevented her from performing arduous physical work within the home. The difference between the sexes in physical strength 'is not now,' she says, 'any more than in the past, a sufficient explanation of woman's relegation to inferior status' (Mitchell 1973 : 105).

It has also been pointed out that the development of machinery – a development which Marx and Engels regarded as contributing to women's emancipation – has been a double-edged weapon as far as the position of women is concerned. The development of machinery did indeed eliminate the biological disadvantage of women in terms of physical strength, but at the same time, this development also made labour more expendable (Mitchell 1973; Saffioti 1978). And because the prevailing 'familial ideology' (Barrett 1980 : 204) defines women's place as primarily in the home (and out of the labour market), women have been an easy target when it comes to laying off workers. This ideology's assumption of women's economic dependence on men has also meant that employers have been able to get away with paying women much lower wages than men. As a result, women's participation in the labour force has not, in the hundred years since Engels wrote *The Origin of the Family*, resulted in the kind of equality between the sexes to which Engels anticipated that it would eventually lead. As Delmar says: 'although women are certainly in production (often on a part-time basis) the female rate of pay is so much lower than the male rate that economic equality is still beyond the horizon' (Delmar 1976 : 287). Lastly, despite the developments of modern industry woman's lesser strength continues to be adduced as a reason why they should be paid less than men.

Many of the above criticisms of Engels are valid. They demonstrate that Engels did not provide a completely adequate account of the causes of the development of patrilineal descent systems, nor of the factors determining the character of the sexual division of labour in human society. Nevertheless these criticisms leave intact Engels's claim that biology – the fact that women bear children, and the fact that they are weaker than men – in combination with changes in the mode of production directly affects, even if it does not entirely explain, women's status in society.

Radical feminism

Whereas liberal feminism and marxism both assert that social factors are prior to biological factors in determining women's social status, many radical feminists have asserted the converse of this, namely that it is not society but biology – women's role in reproduction and men's greater physical strength – that is the primary cause of women's subordinate status. Radical feminism developed explicitly out of a rejection of both liberal feminist and marxist perspectives on the woman question. There is, say radical feminists, a class division, a political division, between men and women – one that many of them believe to be determined by biological sex and that can only be resolved through 'feminist revolution' (Firestone 1974), not through the reformist tactics adopted by liberal feminist organizations such as NOW.

While describing itself as a revolutionary movement, and claiming to draw on the materialist and dialectical philosophy of marxism, radical feminism at the same time rejects marxist analyses of sexual divisions in society. Firestone, for instance, claims that marxism has been bankrupted by the male supremacist attitudes inherent in its theory and practice 'in which, of course, Ladies never go first' and where 'Feminism is secondary in the order of political priorities, and must be tailored to fit into a pre-existent (male-created) political framework' (Firestone 1974:35–6). Women, say many radical feminists, are a class, and it is the biological division of individuals by sex, not the relation of individuals to the means of production, that constitutes the first class division in society. Sex class, and 'the power men everywhere wield over women' (Rich 1980:660) is, they claim, the origin and 'model' of all other divisions and forms of 'exploitation' in society such as those of race and of social class (e.g. Mehrhof 1969).[4] That is, radical feminists often given priority to biology – to the biological distinction of sex – rather than to the mode of production in their analysis of the cause of women's oppression.[5]

Specifically, said the radical feminists of the late 1960s and early 1970s, it is women's role in reproduction that causes their oppression. Women's dependence on and hence exploitation by men is, they said, the direct consequence of the fact that women bear and are therefore responsible for raising children (e.g. Dunbar 1970). Radical feminists thus often agree with anti-feminist biological determinists in viewing women's dependence on men as the effect of a biologically determined relation between childbearing and childrearing.

Radical feminism has, to some extent, now been replaced by 'cultural feminism', by a movement which seeks to celebrate women's 'culture'.[6] This version of radical or revolutionary feminism asserts that it is not so much the biological fact that women bear children as the psychological construction of this fact – men's envy of women's reproductive processes and of 'female creative energy in all of its dimensions' (Daly 1978:60) – that leads men to control and make women subordinate to them (e.g. Rich 1977; Hanmer *et al.* n.d.; Jeffreys n.d.). Both radical and cultural feminists often agree, however, in arguing that woman's biological role in reproduction (either directly, or indirectly via its psychological construction) accounts for her inferior status, and both tend to agree in viewing men as the main enemy of women's liberation.

Thus radical feminism maintains that women's biological role in reproduction is a primary cause of their subordinate status. It also maintains that other biological sex differences – namely sex differences in strength, aggression, and the capacity for violence – reinforce women's oppression; that men use their physical strength to subdue women, and to maintain their dominance over them (e.g. Redstockings 1969; Millett 1971; McKenzie n.d.).

The solutions advocated by radical feminism to the woman question have been various. On the one hand there is Firestone's notorious suggestion that women's oppression should be resolved through artificial reproduction. On the other hand, cultural feminists (e.g. Rich 1977) have recommended that, far from rejecting their childbearing function, women should celebrate it. Like some of the feminists of the nineteenth and early twentieth centuries (see Chapter 3 above), cultural feminists subscribe to the biological essentialist thesis that biology has endowed women with distinctive 'feminine' traits and that these should be valued equally with, if not more than, men's specific psychological traits (e.g. Dunbar 1970; Daly 1978). Roxanne Dunbar, for instance, asserts that the 'female capacity for reproduction' led to women's dependence on men, but that at the same time it led to the development, in women, of certain 'feminine' character traits that should be given greater value and should be the basis for feminist revolution. Maternity, she says,

> 'develops a certain consciousness of care for others, self-reliance, flexibility, non-competitiveness, cooperation and materialism . . . If these "maternal" traits, conditioned into women, are desirable traits, they are desirable for everyone . . . Women and other oppressed peoples must lead and structure the revolutionary movement and the new society to assure the dominance of feminist principles.' (Dunbar 1970:490)

Similarly, Firestone herself argues for the celebration of femininity as one of the goals of radical feminism, for the development of 'a new way of relating . . . one that will eventually reconcile the personal – always the feminine prerogative – with the public, with the 'world outside,' to restore that world to its emotions, and literally to its senses' (Firestone 1974:38).[7]

Others have adopted a more clearly biologically determinist line on this matter. Dunbar argues that 'feminine' traits are an indirect consequence of women's biology, and of the fact that their biological role in childbearing results in girls being socialized towards femininity. Others, however, have accepted recent biological determinist accounts of the origins of 'feminine' traits. Gina (1974), for instance, accepts recent claims regarding the cortical determination of sex role differences (see Chapter 6 above) and argues that masculine and feminine traits are the direct consequence of sex differences in the brain. Although she claims that femininity is directly rather than indirectly determined by biology, she is similar to other radical feminists in putting forward the thesis that greater value should be attached to women's supposed essential femininity: she urges a celebration of women's supposedly distinctive right hemispheric skills.

On the matter of the role of male violence in reinforcing women's subjugation, many radical feminists (e.g. Millett 1971) are agreed that male violence was crucial in determining women's subordination in the first place but that its effect is now tempered by social institutions of control. Some writers (e.g. Brownmiller 1975; Pizzey 1974) go on to argue that we should seek to strengthen the powers of these institutions (e.g. the law, the police) in order to combat the violent abuse of women by men even though, as Alison Edwards (1979) in America points out, such measures may also increase the harassment of other oppressed groups in society, such as the Blacks.

Other radical feminists (e.g. Atkinson 1970; Holliday 1978) regard sex inequality as still determined primarily by men's greater capacity for violence. Laura Holliday, for instance, accepts the arguments of sociobiology that male dominance is the effect of male aggression and that this is ultimately determined by male hormones. She claims that 'Just as dominance is established between primate males by intermittent tests of physical prowess, so the male establishes his power over females by occasional reminders of his brute strength.' (Holliday 1978:131) Since male dominance is, in her view, the effect of biology, it can only be curtailed by factors affecting the biochemistry of aggression – by, she suggests, biofeedback techniques, by living in less crowded and less polluted environments, and by vegetarianism!

Having outlined the radical feminist account of the relation between biology and women's social status, I shall now turn to a critique of it. I have already criticized the sociobiological claim that male dominance is determined by 'male' hormones (see Chapter 5 above). These criticisms apply with equal force to Holliday's statement of it. The more usual radical feminist arguments about male aggression, namely that it has from earliest times been used to reinforce male dominance, is also questionable. The evidence suggests that the earliest forms of human social organization were not based on male dominance (Leacock 1975). There is therefore no good basis for asserting that men's biologically given capacity for aggression made them dominant in these early societies.

The radical feminist assertion that women should seek liberation through gaining a higher valuation of their supposedly essential 'feminine' characteristics is, I have argued, also untenable. In the first place, the claim that women have special feminine traits is based on abstraction not on concrete fact; women simply do not demonstrate most of the traits stereotypically associated with femininity (see Chapter 3 above). Second, such a solution to the woman question is an entirely idealist one. It trades on the futile hope that ideas – that moralism and the revaluation of femininity – can, on their own, change the concrete reality of women's lives (Mitchell 1973; Guettel 1974). Third, as a number of writers have pointed out both from within radical feminism (e.g. Brooke 1978), and from the perspectives of marxism and socialist feminism (e.g. Dixon 1978), the celebration of a distinctive, supposedly biologically given femininity, and of women's 'separate' sphere is reactionary. It may have been progressive and realistic in the past to hope to achieve sexual equality through gaining an equal valuation for women's role in the home with men's role in work outside the home. But the goal of achieving sexual equality on the basis of confining women and men to separate spheres of activity has now been rendered entirely anachronistic and futile by the historical development of the economy over the last several centuries (see Chapter 9 above for a more extensive discussion of this point).

Although Firestone flirts with cultural feminism, her basic analysis of sex roles is materialistic rather than idealist. Nevertheless her biological materialist analysis is flawed by its lack of historical perspective, by the fact that it assumes that women's economic dependence on men has been determined for all time by the biological fact that women bear children. Certainly this fact is, as I have argued repeatedly above, relevant to the explanation of why women came to be economically dependent on men during the course of history. Nevertheless, this biological fact has not always made women dependent on men nor need it do so in the future. Women's dependence on men changes as a result of changes in the mode of production and of the way biological sex differences are articulated with these modes of production. It is appropriate changes in the mode of production, not the kinds of changes in biology recommended by Firestone, that will lead to significant improvements in women's social status relative to men. As Engels demonstrated, women's subordination to men was already being undermined by industrial development as he wrote. He anticipated that it would be still further undermined given the abolition of private property. Women's social subordination has not been given for all time by biology, as some radical feminists would have us believe, nor is it to be changed by simply technologizing women's biological role in reproduction as Firestone advises (Rose and Hanmer 1976).

I shall conclude this section by addressing the radical feminist claim that men are the main enemy of feminism. Certainly, many of the most fervent opponents of women's rights have been men. Nevertheless, an important cause of their complaint against equal rights in both middle- and working-class jobs resides in the fact that in conditions of job scarcity and exploitative wage structures, women's entry into these occupations has often depressed wages and been at the cost of men's jobs. It is not men, but these economic conditions – conditions that oppose the interests of men and women alike – that constitute a basic cause of women's oppression, at least in the labour market. As several critics (e.g. Reed 1970; Stone 1972) of this aspect of radical feminism have pointed out, the view that men are the main enemy of women's liberation is not new within feminism. Nor is it new for men to oppose sexual equality at work in the belief that this is the best way of securing their interests as regards wages and employment. In 1867, for instance, the Lassallean group within the German socialist movement objected, in these terms, to women's entry into the industrial labour force. August Bebel pointed out then the short-sightedness of this attempt, in the name of socialism, to exclude women from the factories. The interests of men and women in the working class would, he said, in the long term be best served by the entry of women into the labour force, by women joining forces with men in the struggle against private capital (Draper and Lipow 1976). Similarly, when men in England criticized women's entry into the work force on the ground that it would undercut the general wage rate, Eleanor Marx (1892) pointed out that it was not women but capital that constituted the main obstacle to their securing adequate wage levels. It was not women who were the main enemy then and it is not men who are the main enemy now. Rather, one of the primary obstacles to women (as well as men) securing good working conditions is capital, which (as we have seen in Chapter 2) often uses inter-sexual competition in the work force to mask this fact.

Socialist feminism

Socialist feminism, like radical feminism, has developed as a result of dissatisfaction with both liberal and classical marxist analyses of sexual inequality. Both socialists (e.g. Rowbotham 1979) and radical feminists (e.g. Firestone 1974) have criticized the organization of left wing groups and the way they have treated campaigns around women's issues – campaigns for nursery provision and abortion and against sexual discrimination at work – as of secondary importance. Unlike radical feminists, however, socialist feminists (e.g. Eisenstein 1978; Kuhn and Wolpe 1978) have not sought to subordinate marxism to feminism. Instead they have attempted to give marxism and feminism, and their accounts of capitalism and sexual inequality, equal weight within a socialist feminist account of sexual divisions in society.

Marxist theory, say many socialist feminists (e.g. Rubin 1975), needs to be supplemented by feminist theory since the former fails to account for the fact that sexism predates capitalism. Others (e.g. Kuhn 1978), however, recognize that, since Engels located the origins of the subordination of women in the early development of herding societies (i.e. in societies which antedated by thousands of years the development of capitalism), classical marxist theory is quite capable of countenancing the fact that sexism predates capitalism. Engels, however, does not explain, they say, 'the persistence of the working-class family within capitalism' (Humphries 1977 : 241); he failed to explain why sexism continues to prevail within the proletariat – within social classes which do not own private property (e.g. McDonough and Harrison 1978). Socialist feminists agree with radical feminists (e.g. Millett 1971 : 121) in pointing out that Engels was unjustifiably romantic in his account of relations between the sexes in the proletarian family in which, he said, lack of property had 'cleared away . . . [all] incentive to make . . . male supremacy effective' (Engels 1972 : 135).

Nevertheless, as some marxist feminists (e.g. Humphries 1981; Gimenez in press) have also pointed out, sexual inequalities in the working class may well reflect – at least in part – the success of the middle classes in enforcing their biologically and property based sexual mores on the non-propertied classes in order to ensure that the sex-role norms of these classes conformed with, and did not undermine middle-class mores. If this is indeed the case then Engels's claim that women's biological role in reproduction, in conjunction with the development of private property, shapes existing divisions between the sexes applies (albeit differently) to both the working class and the middle class.

Perhaps the greatest cause of socialist feminist disappointment in marxism lies in the fact that the traditional unequal division of labour between the sexes within the family continues to prevail in socialist countries (e.g. de Beauvoir 1976; Scott 1979). Some have attempted to explain this in terms of marxist theory (e.g. Delmar 1976; Leacock 1978) – in terms of the dialectical relationship that it posits between the forces and relations of production,[8] and in terms of the relatively low level of economic development in many socialist societies (e.g. Saffioti 1978). The majority of socialist feminists, however, have argued that the continuation of sexism within socialist countries is due to women's specific situation within the family being relatively autonomous from the mode of production. They conclude that Engels was therefore wrong to suppose that this situation would be transformed as a result of the socialization of production (e.g. Zaretsky 1976; Coward 1978a).

Some socialist feminists (e.g. Mitchell 1973; Eisenstein 1978) have attributed this 'mistake' of Engels to the fact that both he and Marx regarded the sexual division of labour within the family as given by biology, and to the fact that they failed to criticize this division. This failure on the part of classical marxism constitutes a particularly serious matter in the eyes of many feminists seeking a marxist analysis of women's situation, since they point out that the family constitutes a particularly important site of women's oppression today. Some have sought to make good this gap in marxism by arguing that woman's place in the family, particularly her domestic labour, can be accounted for in terms of the mode of production (e.g. Harrison 1973) or in terms of Marx's analysis of 'labour power' (e.g. Seccombe 1973). Others, by contrast, reject these particular attempts to accommodate feminism within marxism. They have either sought to develop a more adequate marxist economic analysis of housework (e.g. Himmelweit and Mohun 1977) or they have argued that marxist economics needs to be supplemented with, say, psychoanalysis if we are fully to understand how women's oppression is reproduced within the family (e.g. Mitchell 1974; Rubin 1975; Chodorow 1978).

It is in this second line of approach that we find most frequent reference by socialist feminists to the place of biology in shaping women's oppression. Juliet Mitchell, disappointed in the failure of marxism to analyze or criticize the position of women in the family, has as we have seen (Chapter 8 above) resorted to a Lacanian reading of Freud to fill this apparent gap in marxist theory. Patriarchy, she says, is not the effect of economics nor of specific modes of production; nor, she says, is it determined by biology. Instead, she claims that it is constituted by the exchange of women by men and by the construction of biological sex difference in terms of this exchange, a construction which manifests itself in early childhood in the form of the castration complex.

It is true to say that most socialist feminists 'reject analyses which locate the basis of women's subordination in our biology' (Page 1978:33). At the same time, it is clear that in so far as socialist feminist theory has been influenced by Lacan's account of the castration complex, it has given biology, at least its social construction, an important place in its account of women's social status. For, on the basis of Lacan, these socialist feminists claim that the social construction of biological sex difference both reflects and contributes to the reproduction of patriarchy and existing sexual divisions in society.

I have argued above (Chapter 8) that the Lacanian account of Freud, as used by Mitchell and Coward, misrepresents Freud in ways that do not serve the interests of feminism: that in the first place, it fails to deal adequately with the direct effects of biology on women's lives; and, in the second place, it unwarrantably provides an account of women's destiny which is just as deterministic as that provided by biological determinism. Several writers (e.g. Eisenstein 1978; Beechey 1979) have criticized Mitchell in these terms. Others (e.g. Leacock 1972, Edholm, Harris, and Young 1977) point out that accounts of sexual inequality such as Juliet Mitchell's rest on the mistaken belief that patriarchy and kinship are uniform. It has also been argued by, for instance, Nancy Chodorow (1978) and Michèle Barrett (1980) that Mitchell is too charitable to Freud who, they say, gives too much weight to biology in his account of psycho-sexual differentiation. It is therefore not surprising to find that, although these authors criticize Mitchell, they also favour a social constructionist account of the influence of biology in shaping sexual inequality. In the case of Chodorow this has meant rejecting Freud's biologically based account of infant development (i.e. in terms of oral, anal, and genital stages) in favour of an object relations account of that development (i.e. in terms of stages in the infant's social relations first with its mother, and then with its mother and father).

The psychoanalytically based accounts of sexual inequality provided by writers like Juliet Mitchell, Gayle Rubin, and Nancy Chodorow tend to treat its determinants as relatively autonomous from the mode of production. Some have sought to justify this dualism by reference to Engels's statement that:

> 'According to the materialistic conception, the determining factor in history is, in the final instance, the production and reproduction of immediate life. This, again, is of a twofold character: on the one side, the production of the means of existence, of food, clothing and shelter and the tools necessary for that production; on the other side, the production of human beings themselves, the propagation of the species. The social organization under which the people of a particular historical epoch and a particular country live is determined by both kinds of production: by the stage of development of

> labor on the one hand and of the family on the other.' (Engels 1972:71–2)

This passage in Engels, it is said (e.g. Rubin 1975), demonstrates that he regarded production and reproduction as relatively autonomous activities. Moreover, it is argued (e.g. Mitchell 1973; Coward 1978b) that Althusser's reading of Marx shows that it is quite consistent with marxism to assert that the superstructure, as reflected in patriarchal ideology, is relatively autonomous from the economic base; that it is quite consistent with marxism to reject 'economistic' accounts of sexual inequality. Such readings are, however, a travesty, both of marxist theory in general and of Engels's account of the relation between reproduction and production in particular. As I have sought to show above, Engels quite clearly maintains that the social elaboration of women's biologically given role in reproduction is not autonomous from, but is in fact quite directly related to, the mode of production. In separating reproduction from production, such authors have replaced Engels's materialist account of women's oppression with an idealist one, an account that (in the case of Chodorow 1978) explains sexual inequality in terms of the psychology of the mother-infant relationship, or (in the case of Mitchell 1974) explains it in terms of the ideas that the child entertains about genital sex difference.

A number of writers have pointed out the various ways in which the family and reproduction, far from being autonomous from production, are indissolubly linked with it (e.g. Braverman 1976). They have also pointed out that the treatment of family life as autonomous from production is itself the product of changes in production and of the way in which historical forces have led to the separation of public and private life in the capitalist mode of production (e.g. Zaretsky 1976). As a result, a number of socialist and marxist feminists (e.g. Barrett 1980) have sought to integrate their accounts of women's position in the family more firmly within an historical account of changes in the mode of production.

Several writers (e.g. Zaretsky 1979; Kelly 1979) have signalled their willingness to unify psychoanalytic and marxist perspectives on these two phenomena. There has, however, been a dearth of actual attempts to do so. Some suggest that this is because this project is illusory, that psychoanalysis and marxism are incompatible 'discourses' (Adlam 1979). In the writing of Juliet Mitchell (1974) and

Nancy Chodorow (1978) these discourses do indeed appear to be incompatible. Both these writers use psychoanalysis to develop accounts of sexual inequality that are essentially non-marxist in that they fail to explain how sexual inequality varies historically with changes in the mode of production. But this is less the fault of Freud than of the fact that both these writers assume that the character of sexual inequality is essentially constant across all modes of production.

As I have pointed out above (Chapter 8) Freud recognized the falsity of this premise. Like Engels, he recognized that historical factors alter the dimensions of women's oppression. Thus, although recent socialist feminist uses of psychoanalysis do indeed make psychoanalysis appear incompatible with marxism, Freud and Engels did in fact cover a similar terrain in their accounts of women's social situation, in that they both insisted that biology and history together directly shape women's destiny.

Biological sex and social class

I shall now consider one last way in which most of the theories of contemporary feminism incorporate biology – the fact of biological sex – within their accounts of women's status in society. A common theme within liberal, radical, and socialist feminism has been to assume that because of their shared biological sex women also share a common situation within society, a situation in the family, for instance, which is essentially uniform across different social classes.

Liberal feminists have often assumed that the liberal ideology that represents the interests of middle-class women also necessarily represents the interests of working-class women. On some issues, such as that of equal pay, and in the past that of the suffrage, the interests of both classes do coincide, though it is noteworthy that many of the arguments for the suffrage advanced by the middle classes were couched in terms that appealed to their interests and explicitly opposed those of the working class (Magas 1971: 85–6). On other issues, however, such as protective legislation, the interests of women in the bourgeoisie conflicted with and continue to conflict with those of working-class women. At the turn of the century, for instance, women of the bourgeoisie objected to the introduction of protective legislation because it threatened to provide a wedge for anti-feminists whereby they would be able to argue that since women had shorter working hours than men, they should not therefore have equal rights with them in professional employment. Working-class women, on the other hand, often welcomed protective legislation on the grounds that it restricted the otherwise remorseless way in which industry exploited their cheap labour. They also welcomed such legislation on the grounds that it would promote the physical well-being of women as mothers, and would provide a basis upon which to argue for a similar reduction in the working hours of men (Draper and Lipow 1976).[9]

A similar conflict between the interests of middle- and working-class women arises over the biological fact of menstruation. Liberal feminists argue that menstruation does not impede women's function, and therefore that this aspect of women's biology should not be a consideration in the employment of women. The motivation for this argument in the case of middle-class women is clear, for as we have seen, the handicaps of menstruation have repeatedly been urged against their equal access to higher education (see Chapter 2 above) and to middle-class professions (see Chapter 7 above). Nevertheless, the claim that women are unhampered by menstruation and that menstrual distress can and should be resolved through drug treatment has, in the case of working-class occupations, not served the interests of women. Instead this latter claim has more often been used as a basis for criticizing the provision of rest periods or paid leave during menstruation, and as a basis for opposing these attempts to improve women's working conditions (see Chapter 7 above). It is noteworthy that in China, where the interests of the factory worker are given priority over any abstract doctrine of equal rights, the effects of menstruation are not denied but are instead catered for through the provision of rest periods and, if necessary, of acupuncture.

The clearest statement that biological sex constitutes the basis for unity between women of different social classes comes from radical feminists who assert that women constitute a class in their own right. Just as socialist feminism finds passages in Engels to justify its claim that reproduction is distinct from production, so radical feminism also claims to find support in Engels for its thesis that biology renders women a distinct class. Engels, says Firestone, observed that

'the original division of labour was between man and woman for the purposes of child-breeding; that within the family the husband was the owner, the wife the means of production, the children the labor; and that reproduction of the human species was an important economic system distinct from the means of production.' (Firestone 1974:4–5)

Certainly Engels does claim that with the development of surplus wealth and of patrilineal descent women became for man 'a mere instrument for the production of children', and that within the family the husband 'is the bourgeois, and the wife represents the proletariat' (Engels 1972:121, 137). Engels's claim in this respect does indeed appear to imply that women constitute a class – a class defined by their relation to reproduction rather than to the means of production (Eisenstein 1978). Elsewhere, however, Engels (1972:129) argues quite clearly that the oppression of women by men is distinct from class exploitation; that the coincidence of these two forms of oppression in point of historical origin was due to their both resulting from the same cause, namely that of the development of private property, not to their being identical forms of exploitation.

As I have pointed out, women's interests as a sex are not necessarily the same in different social classes. Nevertheless many socialist feminists, despite their marxism, assert, like the radical feminists, that women (e.g. Ehrenreich 1975), or at least housewives (Zaretsky 1976), constitute a class. They claim that class divisions between women are more apparent than real (e.g. Charlotte Perkins Gilman Chapter 1978). Some of these feminists even argue that it is quite compatible with marxist theory to provide an analysis of women's social status which dislodges social class from any central place in that analysis (e.g. Coward 1978b).

The belief that women are a class – and that since they share the same biology they must therefore be united on all issues which concern women as a sex – has bedevilled socialist approaches to the woman question for the last century (Slaughter 1979). In 1895, for instance, Clara Zetkin pointed out that the editors of the socialist journal *Vorwärts* had been misled by this belief into thinking that, because of their shared sex, middle- and working-class women must therefore also share the same political priorities. As a result, she said, these editors had wrongly subordinated the working women's campaign for the right of assembly to the middle-class women's campaign against prostitution (Draper and Lipow 1976).

Today, although many socialist feminists are similarly misled by the fact of women's shared biology into assuming that women of different classes have identical interests, others (e.g. Rowbotham 1973; McDonough and Harrison 1978) are aware and stress that social class does divide women on many issues. Even as housewives, it is pointed out (e.g. Guettel 1974; Saffioti 1978), women of the middle class have different problems from those of the working class. And this is borne out by the fact that although both working- and middle-class women are often understandably dissatisfied with their roles as housewives (Gavron 1966; Oakley 1975) their dissatisfactions often vary as a function of class. Certainly the degree to which they suffer from this role in terms of mental ill health is markedly affected by social class. Working-class housewives are, for instance, found to be at an enormously higher risk of psychiatric distress than middle-class housewives (Brown, Brolchain, and Harris 1975). Recent attempts (e.g. Kuhn and Wolpe 1978; Eisenstein 1978) to redress the previous neglect of social class in socialist feminist theory are therefore welcome: that socialist feminist analysis be correct on this matter is essential if practice is not to emphasize the interests of middle-class women at the expense of working-class women.

Conclusion

Finally, I hope I have demonstrated both in the book as a whole, and through my account of the place of biology within feminist theory, that the answers to the woman question are not to be found solely in biology as many radical feminists and conservative anti-feminists have claimed. Nor can they be found in the flat denial by liberal feminism that biological factors like menstruation affect women's lives, nor in the claim of some socialist feminists that biology influences women only via its social construction within patriarchy. The account of women's destiny provided by classical marxism, though it is (as I have indicated) incomplete, is nevertheless one of the few theories underlying contemporary feminism that begins to do justice to the fact that biology does have a direct influence on women's lives, and to the fact that, despite their shared biology, women of different classes nevertheless have different, and sometimes conflicting, interests even in regard to their shared biology. Moreover the recent resurgence of feminism has resulted in a number of extremely useful elaborations of the ways in which the marxist theory of social class can and should inform feminist theory and practice (e.g. Saffioti 1978; Gimenez 1978).

The goals and priorities of the women's movement should not be determined on the basis of abstract theory, whether it is the theory of liberalism, radicalism, cultural feminism, or structuralism. These aims should instead be determined by examining the concrete ways in which the majority of women are oppressed within their sexual and familial relations and at work. Only on the basis of this kind of practice combined with adequate theory can the correct objectives for feminism be determined. The women's movement of the last decade has done much to bring attention to the actual and specific nature of women's oppression under capitalism. It has also done a great deal in organizing women around concrete issues, issues which include those of biology. Campaigns around biological issues have, for instance, included struggles designed to achieve better health care for all women; higher standards of safety in contraception; the right to abortion and the facilities that would make this right equally available to all women.

Capitalism, despite its liberal protestations, has had to be fought every inch of the way to get it to concede equal rights to women, and to get it to make real improvements in the condition of women – in, for instance, the family, in health care,[10] and at work. This fight has meant, among other things, that feminists have repeatedly had to address the biological arguments adduced by the ideologues of capitalism. They have repeatedly had to confront and criticize the many biologically phrased arguments designed to defend existing sexual inequalities in society. As I have demonstrated in this book, these arguments are still being advanced today. The struggle is by no means over; even the limited victories that have been secured in regard to women's biological function in the area of abortion, for instance, are now under threat. Given continued struggle against the reactionary forces that seek to undo the gains of the women's movement, and given continued vigilance for, and criticism of, biological and other ideological defences of reactionary social policy, women and men will not only help secure but will also help extend the gains made for women over the last century.

Notes

1 They disagree with each other, however, as to the precise nature of this

relationship. Some (e.g. Dixon 1978) provide a functionalist account of the relation between existing sexual divisions and capitalism. Others (e.g. Nava 1980) argue that these divisions are not only functional to capital but are also functional to men. Yet others question whether these divisions are ideally suited and, in this sense, functional to the needs either of men or of capital. Instead they explain these divisions in historical, rather than functionalist, terms – in terms of the way they have developed out of pre-capitalist and earlier capitalist modes of production and thus became 'embedded in the structure and texture of capitalist social relations' today (Barrett 1980: 254). Although I agree with Barrett's insistence on developing a historical, not a purely functionalist, account of current sexual divisions in society, it seems to me that her argument against functionalism is not entirely sound. Functionalist analyses of sexual divisions state, she says, that these divisions (expecially divisions within the family) persist because they are functional to capital in reproducing its labour force on a generational, day-to-day, and ideological basis. But, argues Barrett, the family is not the cheapest way for capital to ensure the reproduction of its labour force – immigrant labour is a cheaper source of labour supply because the host country does not have to pay for its reproduction. Nevertheless, since immigrants are themselves born and socialized in families and, in practice, only constitute a marginal part of the labour force within modern capitalist production, Barrett's argument hardly disproves the functional utility of the family (and of the sexual divisions predicated on it) to capital.

2 For examples of earlier biological arguments adduced by socialists against feminism see Draper and Lipow (1976), and Slaughter (1979).

3 For an interesting account of this wariness as it manifested itself in Russia in the early decades of this century see Cathy Porter's (1980) biography of Alexandra Kollontai.

4 This viewpoint is not new within feminism. For an earlier statement of it see, for instance, Thompson (1914).

5 This is not true of all radical feminists. Christine Delphy (1977), for instance, argues that it is women's common position in relation to the domestic, as opposed to the industrial, 'mode of production' which constitutes them as a social class. Her argument is, however, flawed by its dependence on an unwarranted extension of Marx's concept of the mode of production (Barrett and McIntosh 1979).

6 Many 'revolutionary' or 'radical' feminists (e.g. Brooke 1978) would dispute the view that cultural feminism grew out of, or is in any way similar to radical feminism. The celebration of femininity, of a 'culture of women' (New York Radical Women 1970: 520), has, however, always been a part of radical feminist theory. It has also been an aspect of some liberal feminist accounts of sexual inequality (e.g. Hoffman 1972; Lewis 1976). Liberal cultural feminism has, however, differed from radical

cultural feminism in locating the determinants of femininity primarily in socialization rather than in biology.

7 The contradictory way in which Firestone both demands the abolition of sex difference and at the same time celebrates it is also apparent in other contemporary writings on sexual divisions in society. Branka Magas (1971) points to a similar contradiction in Germaine Greer's *The Female Eunuch*.

8 I have found Sean Sayers's (1980) account of this aspect of marxist theory particularly helpful here.

9 For an extremely sound statement on a similar debate now taking place in England over the question whether existing legislation limiting women's working hours should be repealed see Coyle (1980). The related issue of protective legislation in respect of reproductive hazards at work is dealt with at greater length in Chapter 2 (above), and in a special issue of the 1979 volume of *Feminist Studies*.

10 For an example of the earlier struggle for better health and maternity care for women see Davies (1978). The struggle continues in this area. Working-class women's access to good maternity services is still well below that of middle-class women. For documentation on this point as regards the poorest section of English society see, for example, Wedge and Prosser (1973).

References

Addams, J. (1907) *Newer Ideals of Peace*. Excerpted in A. Rossi (ed.) *The Feminist Papers* (1974) New York: Bantam Books.

Addams, J. (1922) *Peace and Bread in Time of War*. New York: Macmillan.

Adlam, D. (1979) The Case Against Capitalist Patriarchy. *m/f* **3**: 83–102.

Alexander, S. (1976) Women's Work in Nineteenth-Century London: A Study of the Years 1820–50. In A. Oakley and J. Mitchell (eds) *The Rights and Wrongs of Women*. Harmondsworth: Penguin.

Allan, J. McGrigor (1869) On the Real Differences in the Minds of Men and Women. *Anthropological Society of London, Journal* **7**: 195–219.

Allen, G. (1889) Plain Words on the Woman Question. *Fortnightly Review* **46**: 448–58.

Allen, N. (1870) Physical Degeneracy. *Journal of Psychological Medicine* **4**: 725–64.

Althusser, L. (1971) Freud and Lacan. In *Lenin and Philosophy and Other Essays*. London: New Left Books.

Amos, S. M. (1894) The Evolution of the Daughters. *Contemporary Review* **65**: 515–20.

Anderson, E. G. (1874) Sex in Mind and Education: A Reply. *Fortnightly Review* **15**: 582–94.

Angrist, S. S. (1969) The Study of Sex Roles. *Journal of Social Issues* **25**(1): 215–32.

Annastasi, A. (1968) *Psychological Testing*. New York: Macmillan.

Anonymous (1874) American Women: Their Health and Education. *Westminster Review* **102**: 216–35.

—— (1878) Letter to the *British Medical Journal* (20th April): 590.

Archer, J. (1976) Biological Explanations of Psychological Sex Differences. In B. Lloyd and J. Archer (eds) *Exploring Sex Differences*. London: Academic Press.

Ardrey, R. (1971) *The Territorial Imperative*. New York: Dell.
—— (1976) *The Hunting Hypothesis*. New York: Atheneum.
Aristotle (1912) De Partibus Animalum. In J. A. Smith and W. D. Ross (eds) *The Works of Aristotle*, Vol. V. Oxford: Clarendon Press.
—— (1913) Physiognomonica. In W. D. Ross (ed.) *The Works of Aristotle*, Vol. VI. Oxford: Clarendon Press.
Atkinson, G. (1970) *Radical Feminism*. In S. Firestone and A. Koedt (eds) *Notes from the Second Year*. Cited in C. Guettel (1974).
Bagehot, W. (1872) *Physics and Politics*. London: H. S. King.
—— (1879) Biology and 'Woman's Rights'. *Popular Science Monthly* **14**: 201–13.
Baker, S. W. (1980) Biological Influences on Human Sex and Gender. *Signs* **6**: 80–96.
Banks, J. A. and Banks, O. (1965) *Feminism and Family Planning in Victorian England*. Liverpool: Liverpool University Press.
Barash, D. P. (1977) *Sociobiology and Behavior*. New York: Elsevier.
Barrett, M. (1980) *Women's Oppression Today*. London: Verso.
Barrett, M. and McIntosh, M. (1979) Christine Delphy: Towards a Materialist Feminism? *Feminist Review* **1**: 95–106.
Barry, H., Bacon, M. K., and Child, I. L. (1957) A Cross-Cultural Survey of Some Sex Differences in Socialization. *Journal of Abnormal and Social Psychology* **55**: 327–32.
Beechey, V. (1979) On Patriarchy. *Feminist Review* **3**: 66–82.
Benedek, T. (1959) Parenthood as a Developmental Phase: A Contribution to the Libido Theory. *Journal of the American Psychoanalytic Association* **7**: 389–417.
—— (1970a) The Psychobiology of Pregnancy. In E. J. Anthony and T. Benedek (eds) *Parenthood*. Boston: Little, Brown.
—— (1970b) Motherhood and Nurturing. In E. J. Anthony and T. Benedek (eds) *Parenthood*. Boston: Little, Brown.
Benston, M. (1969) The Political Economy of Women's Liberation. *Monthly Review* **21**(4): 13–27.
Bettelheim, B. (1962) *Symbolic Wounds*. New York: Collier Books.
Bibring, G. L., Dwyer, T. F., Huntingdon, D. S., and Valenstein, A. F. (1961) A Study of the Psychological Processes in Pregnancy and of the Earliest Mother-Child Relationship: I. Some Propositions and Comments. *Psychoanalytic Study of the Child* **16**: 9–24.
Blackwell, A. B. (1875) *The Sexes Throughout Nature*. New York: Putnam's.
Blake, J. (1972) Coercive Pronatalism and American Population Policy. In E. Peck and J. Senderowitz (eds) *Pronatalism: The Myth of Mom and Apple Pie*. New York: Thomas Y. Crowell.
Bleier, R. (1976) Myths of the Biological Inferiority of Women: An Exploration of the Sociology of Biological Research. *University of Michigan Papers in Women's Studies* **2**: 39–63.
Bloch, I. (1909) *The Sexual Life of Our Time*. Translated by M. Eden Paul. London: Heinemann.

Boardman, P. (1978) *The Worlds of Patrick Geddes: Biologist, Town Planner, Re-Educator, Peace-Warrior*. London: Routledge & Kegan Paul.

Boulding, E. (1977) *Women in the Twentieth Century World*. Beverley Hills, California: Sage.

Bowlby, J. (1971) *Attachment*. Harmondsworth: Penguin.

Boyd, R. (1861) Tables of the Weights of the Human Body and Internal Organs in the Sane and Insane of Both Sexes at Various Ages, Arranged from 2614 Post-Mortem Examinations. *Philosophical Transactions of the Royal Society of London* **151**: 241–62.

Braverman, H. (1976) Two Comments. *Monthly Review* **28**(3): 119–24.

Breen, D. (1975) *The Birth of a First Child: Towards an Understanding of Femininity*. London: Tavistock.

—— (1978) The Mother and the Hospital. In S. Lipshitz (ed.) *Tearing the Veil: Essays on Femininity*. London: Routledge & Kegan Paul.

Breines, W., Cerullo, M., and Stacey, J. (1978) Social Biology, Family Studies, and Antifeminist Backlash. *Feminist Studies* **4**(1): 43–67.

Brighton Women and Science Group (1980) *Alice Through the Microscope: The Power of Science over Women's Lives*. London: Virago.

Broca, P. (1862) Cited in J. B. Davis (1868).

—— (1868) On Anthropology. *Anthropological Review* **6**: 35–52.

—— (1873) Reported in *Nature* 8: 152.

—— (1879) Discussion at the Paris Anthropological Society. *Bulletin de la Société d'Anthropologie de Paris* (3 July). Cited in H. Ellis (1929).

Brooke (1978) The Retreat to Cultural Feminism. In Redstockings (eds) *Feminist Revolution*. New York: Random House.

Brooks, W. K. (1879) The Condition of Women from a Zoological Point of View. *Popular Science Monthly* **15**: 145–55, 347–56.

Brown, G. W., Brolchain, M. N., and Harris, T. (1975) Social Class and Psychiatric Disturbance among Women in an Urban Population. *Sociology* **9**: 225–54.

Brown, H. (1910) Woman Suffrage. Paper read before the Ladies' Congressional Club of Washington, D.C. Cited in H. R. Hays (1964).

Brown, J. K. (1971) A Note on the Division of Labor by Sex. *American Anthropologist* **72**: 1073–78.

Brownmiller, S. (1975) *Against Our Will: Men, Women, and Rape*. New York: Simon & Schuster.

Bryden, M. P. (1979) Evidence for Sex-Related Differences in Cerebral Organization. In M. A. Wittig and A. C. Petersen (eds) *Sex-Related Differences in Cognitive Functioning*. New York: Academic Press.

Buchner, L. (1893) The Brain of Women. *New Review* **9**: 166–76.

Buckle, H. T. (1858) On the Influence of Women on the Progress of Knowledge. *Royal Institution Proceedings* **2** (March): 504–05.

Buffery, A. W. H. and Gray, J. A. (1972) Sex Differences in the Development of Spatial and Linguistic Skills. In C. Ounsted and D. C. Taylor (eds) *Gender Differences: Their Ontogeny and Significance*. Edinburgh: Churchill Livingstone.

Bulley, A. (1890) The Political Evolution of Women. *Westminster Review* **134**: 1–8.

Burniston, S., Mort, F., and Weedon, C. (1978) Psychoanalysis and the Cultural Acquisition of Sexuality and Subjectivity. In Women's Studies Group (eds) *Women Take Issue*. London: Hutchinson.

Burrow, J. (1968) Introduction to C. Darwin, *The Origin of Species*. Harmondsworth: Penguin.

Caplan, G. (1961) *An Approach to Community Mental Health*. London: Tavistock.

Carlson, E. and Carlson, R. (1960) Male and Female Subjects in Personality Research. *Journal of Personality and Social Psychology* **61**: 482–83.

Cavell, M. (1974) Since 1924: Toward a New Psychology of Women. In J. Strouse (ed.) *Women and Analysis*. New York: Grossman.

Chamberlain, A. F. (1892) Review of *Per l'Educazione e la Colture della Donna*, by Professor Sergi (1892). *Pedagogical Review* **1**: 532–34.

Charlotte Perkins Gilman Chapter, New American Movement, North Carolina (1978) A View of Socialist Feminism. In A. M. Jaggar and P. R. Struhl (eds) *Feminist Frameworks*. New York: McGraw Hill.

Chasin, B. (1977) Sociobiology: A Sexist Synthesis. *Science for the People* **9**(3): 27–31.

Chavkin, W. (1979) Occupational Hazards to Reproduction: A Review Essay and Annotated Bibliography. *Feminist Studies* **5**(2): 310–25.

Chetwynd, J. (1975) The Effects of Sex Bias in Psychological Research. Paper given at the Annual Conference of the British Psychological Society.

Chinchilla, N. S. (1980) *Ideologies of Feminism: Liberal, Radical, Marxist*. Social Sciences Research Reports, 61. University of California at Irvine.

Chodorow, N. (1978) *The Reproduction of Mothering*. Berkeley: University of California Press.

Claiborne, R. (1974) *God or Beast: Evolution and Human Nature*. New York: Norton.

Clarke, E. H. (1873) *Sex in Education: or, A Fair Chance for Girls*. Boston: J. R. Osgood.

Cleland, J. (1870) An Inquiry into the Variations of the Human Skull, Particularly in the Antero-Posterior Direction. *Philosophical Transactions of the Royal Society of London* **160**: 117–74.

Coltheart, M. (1975) Sex and Learning Differences. *New Behaviour* **1**: 54–7.

Commons, J. R. (1899) A Sociological View of Sovereignty. *American Journal of Sociology* **5**: 1–15.

Conway, J. (1970) Stereotypes of Femininity in a Theory of Sexual Evolution. *Victorian Studies* **14**: 47–62.

—— (1974) Perspectives on the History of Women's Education in the U.S. *History of Education Quarterly* **14**: 1–12.

Cope, E. D. (1888) The Relationship of the Sexes to Government. *Popular Science Monthly* **33**: 721–30.

Cope, E. D. and Kingsley, J. S. (1895) Editor's Table. *American Naturalist* **29**: 825–27.

Coward, R. (1978a) Re-Reading Freud: The Making of the Feminine. *Spare Rib* (May): 43–6.

—— (1978b) Rethinking Marxism. *m/f* **2**: 85–96.

Coward, R., Lipshitz, S., and Cowie, E. (1976) Psychoanalysis and Patriarchal Structures. In *Papers on Patriarchy*. Lewes: Women's Publishing Collective.

Cowles, T. (1936) Malthus, Darwin, and Bagehot: A Study in the Transference of a Concept. *Isis* **26**: 341–48.

Coyle, A. (1980) The Protection Racket? *Feminist Review* **4**: 1–12.

Crichton-Browne, J. (1879) On the Weight of the Brain and its Component Parts in the Insane. *Brain* **1**: 504–18; and **2**: 42–67.

—— (1892) Sex in Education. *Educational Review* **4**: 164–78.

Crook, J. H. (1972) Sexual Selection, Dimorphism, and Social Organization in the Primates. In B. Campbell (ed.) *Sexual Selection and the Descent of Man 1871–1971*. Chicago: Aldine.

Culpepper, E. E. (1979) Exploring Menstrual Attitudes. In R. Hubbard, M. S. Henifin, and B. Fried (eds) *Women Look at Biology Looking at Women*. Boston: G. K. Hall.

Cunningham, D. (1892) Contribution to the Surface Anatomy of the Cerebral Hemispheres. *Cunningham Memoirs of the Royal Irish Academy*, No. 7. Cited in H. Ellis (1929).

Dalton, K. (1969) *The Menstrual Cycle*. New York: Pantheon.

—— (1979) *Once a Month*. Pomona, California: Hunter House.

Daly, Martin and Wilson, Margo (1978) *Sex, Evolution and Behavior*. Belmont, California: Duxbury Press.

—— (1979) Sex and Strategy. *New Scientist* **81**: 15–17.

Daly, Mary (1978) *Gyn/ecology: The Metaethics of Radical Feminism*. Boston: Beacon Press.

Darwin, C. (1896) *The Descent of Man and Selection in Relation to Sex*. New York: Appleton. Original edition, 1871.

—— (1968) *The Origin of Species*. Harmondsworth: Penguin. Original edition, 1859.

Darwin, F. (1888) *The Life and Letters of Charles Darwin Including an Autobiographical Chapter*. London: John Murray.

Davies, M. L. (1978) *Maternity: Letters from Working Women*. New York: Norton. Original edition, 1915.

Davis, J. B. (1868) Contributions Towards Determining the Weight of the Brain in Different Races of Man. *Philosophical Transactions of the Royal Society of London* **158**: 505–27.

Dawkins, R. (1976) *The Selfish Gene*. New York: Oxford University Press.

De Beauvoir, S. (1965) *Force of Circumstances*. Translated by R. Howard. New York: Putnam's.

—— (1976) *The Second Sex*. Translated by H. M. Parshley. New York: Knopf. Original edition, 1949.

De Rios, M. D. (1978) Why Women Don't Hunt: An Anthropologist Looks at the Origin of the Sexual Division of Labour in Society. *Women's Studies* **5**: 241–47.

Deckard, B. S. (1975) *The Women's Movement: Political, Socioeconomic and Psychological Issues*. New York: Harper & Row.

Delaney, J., Lupton, M. J., and Toth, E. (1976) *The Curse: A Cultural History of Menstruation*. New York: E. P. Dutton.

Delauney, G. (1881) Equality and Inequality in Sex. *Popular Science Monthly* **20**: 184–92.

Delmar, R. (1976) Looking Again at Engels's *Origin of the Family, Private Property and the State*. In J. Mitchell and A. Oakley (eds) *The Rights and Wrongs of Women*. Harmondsworth: Penguin.

Delphy, C. (1977) *The Main Enemy*. London: Women's Research and Resources Centre.

Deutsch, H. (1945) *The Psychology of Women: A Psychoanalytic Interpretation*, Vol. II. New York: Grune & Stratton.

Dewey, J. (1886) Health and Sex in Higher Education. *Popular Science Monthly* **28**: 606–14.

Dinnerstein, D. (1976) *The Mermaid and the Minotaur*. New York: Harper & Row.

Distant, W. L. (1874) On the Mental Differences Between the Sexes. *Journal of the Royal Anthropological Institute of Great Britain and Ireland* **4**: 78–87.

Dixon, M. (1972) Ideology, Class, and Liberation. In M. Anderson (ed.) *Mother was not a Person*. Montreal: Black Rose Books.

—— (1978) *Women in Class Struggle*. San Francisco: Synthesis Publications.

Dollard, J., Miller, N. E., Doob, L. W., Mowrer, O. H., and Sears, R. S. (1939) *Frustration and Aggression*. London: Oxford University Press.

Donelson, E. and Gullahorn, J. E. (1977) Psychobiological Foundations of Sex-Typed Behavior. In E. Donelson and J. Gullahorn (eds) *Women: A Psychological Perspective*. New York: Wiley.

Douglas, M. (1966) *Purity and Danger*. New York: Praeger.

Draper, H. and Lipow, A. G. (1976) Marxist Women Versus Bourgeois Feminism. In R. Miliband and J. Saville (eds) *The Socialist Register*. London: Merlin Press.

Dunbar, R. (1970) Female Liberation as the Basis for Social Revolution. In R. Morgan (ed.) *Sisterhood is Powerful*. New York: Random House.

Dweck, C. S. (1978) Achievement. In M. E. Lamb (ed.) *Social and Personality Development*. New York: Holt, Rinehart & Winston.

Easlea, B. (1981) *Science and Sexual Oppression: Patriarchy's Confrontation with Woman and Nature*. London: Weidenfeld & Nicolson.

East Bay Science for the People (1980) Danger: Women's Work. *Science for the People* **12**(2): 6–12.

Eastman, C. (1920) Now We Can Begin. *The Liberator* **3**: 23–4. Reprinted in J. Sochen (ed.) *The New Feminism in Twentieth-Century America*. Lexington, Mass.: D. C. Heath.

Edholm, F., Harris, O., and Young, K. (1977) Conceptualising Women. *Critique of Anthropology* 3(9 and 10): 101–30.

Edwards, A. (1979) *Rape, Racism, and the White Women's Movement: An Answer to Susan Brownmiller*. Chicago: Sojourner Truth Organization.

Ehrenreich, B. (1975) Speech at Socialist-Feminist Conference. *Socialist Revolution* 5(4): 85–93.

Ehrenreich, B. and English, D. (1973) *Complaints and Disorders: The Sexual Politics of Sickness*. Old Westbury, New York: Feminist Press.

—— (1979) *For her own Good: 150 Years of the Experts' Advice to Women*. New York: Anchor Books.

Ehrlich, C. (1975) Evolutionism and the Place of Women in the United States, 1885–1900. In R. Rohrlich-Leavitt (ed.) *Women Cross-Culturally*. The Hague: Mouton.

Eimerl, S. and DeVore, I. (1965) *The Primates*. New York: Time, Inc.

Eisenstein, Z. (1978) Developing a Theory of Capitalist Patriarchy and Socialist Feminism. In Z. Eisenstein (ed.) *Capitalist Patriarchy and the Case for Socialist Feminism*. New York: Monthly Review Press.

Ellis, H. (1929) *Man and Woman: A Study of Secondary and Tertiary Sexual Characters*. Boston: Houghton Mifflin.

Engelmann, G. J. (1900) The American Girl of To-Day. President's Address, American Gynecological Society, Washington. Cited in G. S. Hall (1905).

Engels, F. (1875) Letter to Lavrov of 12 November. In R. L. Meek (ed.) *Marx and Engels on the Population Bomb* (1971) Berkeley: Ramparts Press.

—— (1880) Socialism: Utopian and Scientific. In L. S. Feuer (ed.) (1959) *Marx and Engels: Basic Writings on Politics and Philosophy*. New York: Doubleday.

—— (1972) *The Origin of the Family, Private Property and the State*. New York: International Publishers. Original edition, 1884.

—— (1976) *Anti-Dühring*. Peking: Foreign Language Press. Original edition, 1878.

Etaugh, C. and Spandikow, D. B. (1979) Attention to Sex in Psychological Research as Related to Journal Policy and Author Sex. *Psychology of Women Quarterly* 4: 175–84.

Fairchild, H. P. (1925) *Immigration: A World Movement and its American Significance*. New York: Macmillan.

Fay, M. A. (1980) Marx and Darwin: A Literary Detective Story. *Monthly Review* 31(10): 40–57.

Fee, E. (1974) The Sexual Politics of Victorian Social Anthropology. In M. Hartmann and L. W. Banner (eds) *Clio's Consciousness Raised*. New York: Harper & Row.

—— (1976) Science and the Woman Problem: Historical Perspectives. In M. S. Teitelbaum (ed.) *Sex Differences*. New York: Anchor Books.

—— (1979) Nineteenth-Century Craniology: The Study of the Female Skull. *Bulletin of the History of Medicine* 53: 415–33.

Fernseed, F. (1881) Sexual Distinctions and Resemblances. *Journal of Science* 3 (Series 3): 741–44.

Ferrero, G. (1894) The Problem of Woman, from a Bio-Sociological Point of

View. *The Monist* 4 : 261–74.

Figes, E. (1970) *Patriarchal Attitudes*. London: Faber & Faber.

Filner, R. E. (1980) Science and Marxism in England, 1930–1945. *Science and Nature* 3 : 60–69.

Finck, H. T. (1887) *Romantic Love and Personal Beauty*. London: Macmillan.

Finot, J. (1913) *Problems of the Sexes*. Translated by Mary J. Safford. New York: Putnam's.

Firestone, S. (1974) *The Dialectic of Sex*. New York: Morrow.

Flerx, V. C., Fidler, D. S., and Rogers, R. W. (1976) Sex Role Stereotypes: Developmental Aspects and Early Intervention. *Child Development* 47 : 998–1007.

Fouillée, A. J. E. (1895) *Tempérament et Caractère selon les Individus, les Sexes et les Races*. Paris: Alcan.

Freeman, J. (1971) The Social Construction of the Second Sex. In M. Garskof (ed.) *Roles Women Play*. Belmont, California: Brooks/Cole.

—— (1975) The Women's Liberation Movement: Its Origins, Structures, Impact and Ideas. In J. Freeman (ed.) *Women: A Feminist Perspective*. Palo Alto, California: Mayfield Publishing Co.

Freud, S. (1905) Three Essays on the Theory of Sexuality. *Standard Edition of the Complete Psychological Works of Sigmund Freud*. Vol. 7. London: Hogarth Press and the Institute of Psychoanalysis.

—— (1980a) 'Civilized' Sexual Morality and Modern Nervous Illness. *Standard Edition*. Vol. 9.

—— (1980b) On the Sexual Theories of Children. *Standard Edition*. Vol. 9.

—— (1916–17) Introductory Lectures on Psycho-Analysis. *Standard Edition*. Vol. 16.

—— (1918) The Taboo of Virginity. *Standard Edition*. Vol. 11.

—— (1920a) Beyond the Pleasure Principle. *Standard Edition*. Vol. 18.

—— (1920b) The Psychogenesis of a Case of Homosexuality in a Woman. *Standard Edition*. Vol. 18.

—— (1922) Psycho-Analysis. *Standard Edition*. Vol. 18.

—— (1923) The Infantile Genital Organization: An Interpolation into the Theory of Sexuality. *Standard Edition*. Vol. 19.

—— (1924) The Dissolution of the Oedipus Complex. *Standard Edition*. Vol. 19.

—— (1925) Some Psychical Consequences of the Anatomical Distinction Between the Sexes. *Standard Edition*. Vol. 19.

—— (1930) Civilization and its Discontents. *Standard Edition*. Vol. 21.

—— (1931) Female Sexuality. *Standard Edition*. Vol. 21.

—— (1933) New Introductory Lectures on Psycho-Analysis. *Standard Edition*. Vol. 22.

—— (1935) Letter to Carl Muller-Braunschweig, 21 July. *Psychiatry* (1971) 34 : 329.

Friedan, B. (1965) *The Feminine Mystique*. Harmondsworth: Penguin.

Friedl, E. (1975) *Women and Men: An Anthropologist's View*. New York: Holt, Rinehart & Winston.

Frieze, I., Parsons, J., Johnson, P., Ruble, D., and Zellman, G. (1978) *Women and Sex Roles*. New York: Norton.

Fritchey, C. (1970) The Good Doctor. *New York Post* (3rd August): 25.

Gall, F. J. (1835) *On the Functions of the Brain*, Vol. 2. Translated by Winslow Lewis. Boston: Marsh, Capen & Lyon.

Gardener, H. H. (1887) Sex and Brain Weight. *Popular Science Monthly* **31**: 266–69.

Gavron, H. (1966) *The Captive Wife*. London: Routledge & Kegan Paul.

Geddes, P. and Thomson, J. A. (1890) *The Evolution of Sex*. New York: Humboldt Publishing Co. (Humboldt Library of Popular Science, Nos. 132 and 133).

Gilman, C. P. (1966) *Women and Economics*. New York: Harper & Row. Original edition, 1898.

Gimenez, M. (1975) Marxism and Feminism. *Frontiers: A Journal of Women Studies* **1**(1): 61–80.

—— (1978) Structuralist Marxism on 'The Women Question'. *Science and Society* **42**(3): 301–23.

—— (In press) The Oppression of Women: A Structuralist Marxist View. In I. Rossi (ed.) *Structural Sociology: Theoretical Perspectives and Substantive Analyses*.

Gina (1974) Rosy Rightbrain's Excorcism/Invocation. *Amazon Quarterly* **2**(4): 12–19.

Glennon, L. M. (1979) *Women and Dualism*. New York: Longman.

Goldberg, S. (1977) *The Inevitability of Patriarchy*. London: Temple Smith.

Goleman, D. (1978) Special Abilities of the Sexes: Do They Begin in the Brain? *Psychology Today* **12**(6): 48–59, 120.

Gordon, L. (1976) *Woman's Body, Woman's Right: A Social History of Birth Control in America*. New York: Grossman.

Gould, S. J. (1977) *Ever Since Darwin: Reflections in Natural History*. New York: Norton.

—— (1978) Women's Brains. *New Scientist* **80**(1127): 364–66.

Gray, J. A. and Buffery, A. W. H. (1971) Sex Differences in Emotional and Cognitive Behaviour in Mammals Including Man: Adaptive and Neural Bases. *Acta Psychologica* **35**: 89–111.

Greer, G. (1970) *The Female Eunuch*. London: MacGibbon & Kee.

Griffiths, D. and Saraga, E. (1979) Sex Differences in Cognitive Abilities: A Sterile Field of Enquiry? In O. Hartnett, G. Boden, and M. Fuller (eds) *Sex-Role Stereotyping*. London: Tavistock.

Gross, C. G. (1974) Biology and Pop-Biology: Sex and Sexism. In E. Tobach, J. Gianutsos, H. R. Topoff, and C. G. Gross, *The Four Horsemen: Racism, Sexism, Militarism and Social Darwinism*. New York: Behavioral Publications.

Guettel, C. (1974) *Marxism and Feminism*. Toronto: Canadian Women's Educational Press.

Gwynn, S. (1898) Bachelor Women. *Contemporary Review* **73**: 866–75.

Hall, G. S. (1905) *Adolescence: Its Psychology and its Relations to Physiology, Anthropology, Sociology, Sex, Crime, Religion and Education*, Vol. II. New York: Appleton.

—— (1906) The Question of Coeducation. *Munsey's Magazine* **34**: 588–92.

Haller, J. S. and Haller, R. M. (1974) *The Physician and Sexuality in Victorian America*. Urbana, Illinois: University of Illinois Press.

Hamilton, W. D. (1964) The Genetical Evolution of Social Behavior. *Journal of Theoretical Biology* **7**: 1–52.

Hammond, W. A. (1887a) Brain-Forcing in Childhood. *Popular Science Monthly* **30**: 721–32.

—— (1887b) Men's and Women's Brains. *Popular Science Monthly* **31**: 554–58.

Hanmer, J., Lunn, K., Jeffreys, S., and McNeill, S. (n.d.) Sex Class – Why is it Important to Call Women a Class? *Scarlet Women* **5**.

Hardaker, M. A. (1882) Science and the Woman Question. *Popular Science Monthly* **20**: 577–84.

Harris, B. J. (1978) *Beyond her Sphere: Women and the Professions in American History*. Westport, Connecticut: Greenwood Press.

Harris, L. J. (1977) Sex Differences in the Growth and Use of Language. In E. Donelson and J. Gullahorn (eds) *Women: A Psychological Perspective*. New York: Wiley.

—— (1978) Sex Differences in Spatial Ability: Possible Environmental, Genetic and Neurological Factors. In M. Kinsbourne (ed.) *Hemispheric Asymmetries of Function*. Cambridge: Cambridge University Press.

Harrison, J. (1973) The Political Economy of Housework. *Bulletin of the Conference of Socialist Economists* **3**(1).

Hayes, A. (1891) Health of Woman Students in England. *Education* **11**: 284–93. Cited in G. S. Hall (1905).

Hays, H. R. (1964) *The Dangerous Sex: The Myth of Feminine Evil*. New York: Putnam's.

Herron, J., Galin, D., Johnstone, J., and Ornstein, R. E. (1979) Cerebral Specialization, Writing Posture, and Motor Control of Writing in Left-Handers. *Science* **205**: 1285–89.

Higginson, T. W. (1874) In J. W. Howe (ed.) *Sex and Education: A Reply to Dr E. H. Clarke's 'Sex in Education'*. Boston: Roberts Brothers.

Himmelweit, S. and Mohun, S. (1977) Domestic Labour and Capital. *Cambridge Journal of Economics* **1**: 15–31.

Ho, M-W. and Saunders, P. T. (1981) Adaptation and Natural Selection: Mechanism and Teleology. In M. Barker, L. Birke, A. Muir, and S. P. R. Rose (eds) *Against Biological Determination: Towards a Liberatory Biology*. London: Allison & Busby.

Hobbes, T. (1950) *Leviathan*. London: J. M. Dent. Original edition, 1651.

Hoffman, L. W. (1972) Early Childhood Experiences and Women's Achievement Motives. *Journal of Social Issues* **28**(2): 129–55.

Hofstadter, R. (1955) *Social Darwinism in American Thought*. Boston: Beacon Press.

Holliday, L. (1978) *The Violent Sex: Male Psychobiology and the Evolution of Consciousness*. Guerneville, California: Bluestocking Books.

Hollingworth, L. (1914) *Functional Periodicity: An Experimental Study of the Mental and Motor Abilities of Women During Menstruation*. New York: Teachers College, Columbia University.

Hollis, P. (1979) *Women in Public: The Women's Movement, 1850–1900*. London: Allen & Unwin.

Horney, K. (1926) The Flight From Womanhood. Reprinted in J. B. Miller (ed.) *Psychoanalysis and Women* (1973) Harmondsworth: Penguin.

—— (1932) The Dread of Woman. *International Journal of Psychoanalysis* **13**: 348–60.

Howe, J. W. (1874) In J. W. Howe (ed.) *Sex and Education: A Reply to Dr E. H. Clarke's 'Sex in Education'*. Boston: Roberts Brothers.

Hubbard, R., Henifin, M. S., and Fried, B. (1979) *Women Look at Biology Looking at Women*. Boston: G. K. Hall.

Humphries, J. (1977) Class Struggle and the Persistence of the Working-Class Family. *Cambridge Journal of Economics* **1**(3): 241–58.

—— (1981) Protective Legislation, the Capitalist State, and Working Class Men: The Case of the 1842 Mines Regulation Act. *Feminist Review* **7**: 1–33.

Huschke, E. (1854) *Skull, Brain, and Mind in Men and Animals*. Jena: F. Mauke.

Hutt, C. (1972) *Males and Females*. Harmondsworth: Penguin.

Hyatt, A. (1897) The Influence of Woman in the Evolution of the Human Race. *Natural Science* **11**: 89–93.

Ingleby, J. D. (1980) 'The Politics of Psychology: Review of a Decade'. Unpublished manuscript.

Irigaray, L. (1977) Women's Exile. *Ideology and Consciousness* **1**: 62–76.

—— (1980) When our Lips Speak Together. *Signs* **6**: 69–79.

J.D. (1894) Psychological Literature. *Psychological Review* **1**: 400–11.

Jacklin, C. N. (1979) Epilogue. In M. A. Wittig and A. C. Petersen (eds) *Sex-Related Differences in Cognitive Functioning*. New York: Academic Press.

Jackson, M. B. (1874) In J. W. Howe (ed.) *Sex and Education: A Reply to Dr E. H. Clarke's 'Sex in Education'*. Boston: Roberts Brothers.

Jacobi, M. P. (1877) *The Question of Rest for Women During Menstruation*. New York: Putnam's.

Jaggar, A. M. and Struhl, P. R. (1978) *Feminist Frameworks*. New York: McGraw-Hill.

Jardine, A. (1979) Interview with Simone de Beauvoir. *Signs* **5**(2): 224–36.

Jeffreys, S. (n.d.) The Need for Revolutionary Feminism. *Scarlet Women* **5**.

Jones, E. (1932) The Phallic Phase. In *Papers on Psychoanalysis* (1961). Boston: Beacon Press.

—— (1935) Early Female Sexuality. In *Papers on Psychoanalysis* (1961). Boston: Beacon Press.

Kelly, J. (1979) The Doubled Vision of Feminist Theory: A Postscript to the 'Women and Power' Conference. *Feminist Studies* **5**: 216–27.

Kestenberg, J. S. (1956) On the Development of Maternal Feelings in Early Childhood: Observations and Reflections. *Psychoanalytic Study of the Child* **11**: 257–91.

Kipnis, D. M. (1976) Intelligence, Occupational Status, and Achievement Orientation. In B. Lloyd and J. Archer (eds) *Exploring Sex Differences*. London: Academic Press.

Klein, M. (1924) An Obsessional Neurosis in a Six-Year Old Girl. In *The Psycho-Analysis of Children*. London: Hogarth Press (1975).

Klein, V. (1971) *The Feminine Character*. London: Routledge & Kegan Paul.

Kohlberg, L. and Ullian, D. Z. (1974) Stages in the Development of Psychosexual Concepts and Attitudes. In R. C. Friedman, R. M. Richart, and R. L. Van De Wiele (eds) *Sex Differences in Behavior*. New York: Wiley.

Kraditor, A. S. (1965) *The Ideas of the Woman Suffrage Movement, 1890–1920*. New York: Columbia University Press.

Kreutz, L. E. and Rose, R. M. (1972) Assessment of Aggressive Behavior and Plasma Testosterone in a Young Criminal Population. *Psychosomatic Medicine* **34**: 321–32.

Kuhn, A. (1978) Structures of Patriarchy and Capital in the Family. In A. Kuhn and A. Wolpe (eds) *Feminism and Materialism*. London: Routledge & Kegan Paul.

Kuhn, A. and Wolpe, A. (1978) Feminism and Materialism. In A. Kuhn and A. Wolpe (eds) *Feminism and Materialism*. London: Routledge & Kegan Paul.

Lambert, H. H. (1978) Biology and Equality: A Perspective on Sex Differences. *Signs* **4**(1): 97–117.

Lamott, K. (1977) Why Men and Women Think Differently. *Horizon* **19**(3): 41–5.

Langley, R. and Levy, R. C. (1977) *Wife Beating: The Silent Crisis*. New York: Pocket Books.

Lansdell, H. (1961) The Effect of Neurosurgery on a Test of Proverbs. *American Psychologist* **16**: 448.

—— (1962) A Sex Difference in Effect of Temporal-Lobe Neurosurgery on Design Preference. *Nature* **194**: 852–54.

Lasch, C. (1981) The Transition from Patriarchal Authority to Bureaucratic Authority in Western Europe and America: 5. The Crisis of Nurture. Freud Memorial Lecture in Psychoanalysis, University College, London.

Leacock, E. (1972) Introduction to F. Engels, *The Origin of the Family, Private Property and the State*. New York: International Publishers.

—— (1975) Class, Commodity, and the Status of Women. In R. Rohrlich-Leavitt (ed.) *Women Cross-Culturally*. The Hague: Mouton.

—— (1978) Introduction to H. I. B. Saffioti, *Women in Class Society*. New York: Monthly Review Press.

Le Bon, G. (1878) Recherches Expérimentales sur les Variations de Volume du Crâne et sur les Applications de la Méthode Graphique à la Solution de

Divers Problèmes Anthropologiques. *Comptes Rendus Hebdomadaires des Séances de l'Académie des Sciences* **87**: 79–81.

Lee, A. (1901) Data for the Problem of Evolution in Man – a First Study of the Correlation of the Human Skull. *Philosophical Transactions* **196A**: 225–64.

Leibowitz, L. (1975) Perspectives on the Evolution of Sex Differences. In R. Reiter (ed.) *Toward an Anthropology of Women*. New York: Monthly Review Press.

Lemay, H. R. (1978) Some Thirteenth and Fourteenth Century Lectures on Female Sexuality. *International Journal of Women's Studies* **1**: 391–400.

Lenin, V. I. (1969) *What is to be Done? Burning Questions of our Movement*. New York: International Publishers. Originally published, 1902.

Levi Strauss, C. (1956) The Family. In H. L. Shapiro (ed.) *Man, Culture and Society*. London: Oxford University Press.

Levy, J. (1969) Possible Basis for the Evolution of Lateral Specialization of the Human Brain. *Nature* **224**: 614–15.

—— (1972) Lateral Specialization of the Human Brain: Behavioral Manifestations and Possible Evolutionary Basis. In J. A. Kiger (ed.) *The Biology of Behavior*. Corvallis, Oregon: Oregon University Press.

Levy, J. and Reid, M. (1978) Variations in Cerebral Organization as a Function of Handedness, Hand Posture in Writing, and Sex. *Journal of Experimental Psychology* **107**(2): 119–44.

Lewis, H. B. (1976) *Psychic War in Men and Women*. New York: New York University Press.

Lewis, M. and Weinraub, M. (1979) Origins of Early Sex-Role Development. *Sex Roles* **5**: 135–53.

Lincoln, E. A. (1927) *Sex Differences in the Growth of American School Children*. Baltimore: Warwick & York.

Lloyd, B. B. (1976) Social Responsibility and Research on Sex Differences. In B. Lloyd and J. Archer (eds) *Exploring Sex Differences*. London: Academic Press.

Lopata, H. Z. and Thorne, B. (1978) On the Term 'Sex Roles'. *Signs* **3**(3): 718–21.

Lorenz, K. (1969) *On Aggression*. New York: Bantam.

Lowe, M. (1978) Sociobiology and Sex Differences. *Signs* **4**(1): 118–25.

Maccoby, E. E. and Jacklin, C. N. (1974) *The Psychology of Sex Differences*. Stanford: Stanford University Press.

McCollum, E. V., Orent-Keiles, E. and Day, H. G. (1939) *The Newer Knowledge of Nutrition*. New York: Macmillan. Cited in V. Klein (1971).

MacCormack, C. P. (1977) Biological Events and Cultural Control. *Signs* **3**(1): 93–100.

McDonough, R. and Harrison, R. (1978) Patriarchy and Relations of Production. In A. Kuhn and A. Wolpe (eds) *Feminism and Materialism*. London: Routledge & Kegan Paul.

McDougall, W. (1921) *An Introduction to Social Psychology*. Boston: Luce. Original edition, 1908.

McGhee, P. E. and Frueh, T. (1980) Television Viewing and the Learning of Sex-Role Stereotypes. *Sex Roles* **6**(2): 179–88.

McGlone, J. and Davidson, W. (1973) The Relation Between Cerebral Speech Laterality and Spatial Ability with Special Reference to Sex and Hand Preference. *Neuropsychologia* **11**: 105–13.

McGuinness, D. and Pribram, K. (1979) The Origins of Sensory Bias in the Development of Gender Differences in Perception and Cognition. In M. Bortner (ed.) *Cognitive Growth and Development – Essays in Honor of Herbert G Birch*. New York: Brunner/Mazel.

McIntosh, M. (1981) Feminism and Social Policy. *Critical Social Policy* **1**(1): 32–42.

Macintyre, S. (1977) Childbirth: The Myth of the Golden Age. *World Medicine* **12**(18): 18, 20, 22.

MacKenzie, D. (1981) Notes on the Science and Social Relations Debate. *Capital and Class* **14**: 47–60.

McKenzie, F. (n.d.) Feminism and Socialism. *Scarlet Women* **5**.

McLennan, J. (1865) *Primitive Marriage: An Inquiry into the Origin of the Form of Capture in Marriage Ceremonies*. Edinburgh: A. & C. Black.

Macpherson, C. B. (1962) *The Political Theory of Possessive Individualism: Hobbes to Locke*. Oxford: Oxford University Press.

Magas, B. (1971) Sex Politics: Class Politics. *New Left Review* **66**: 69–91.

Malinowski, B. (1963) *The Family Among the Australian Aborigines: A Social Study*. New York: Schocken Books. Original edition, 1913.

Mall, F. P. (1909) On Several Anatomical Characteristics of the Human Head said to Vary According to Race and Sex, with Especial Reference to the Weight of the Frontal Lobe. *American Journal of Anatomy* **9**: 1–32.

Malthus, T. R. (1960) *On Population*. New York: Random House.

Marx, E. (1892) Letter of 20th May. Reprinted in H. Draper and A. G. Lipow (1976).

Marx, K. (1967) *Capital*, Vol. I. Translated by S. Moore and E. Aveling. New York: International Publishers. Original edition, 1867.

Maudsley, H. (1874) Sex in Mind and in Education. *Fortnightly Review* **15**: 466–83.

Mehrhof, B. (1969) *Class Structure in the Women's Movement*. (*Mimeograph, New York: The Feminists*). Cited in G. G. Yates (1975).

Mill, J. S. (1869) *The Subjection of Women*. Philadelphia: J.B. Lippincott.

Millett, K. (1971) *Sexual Politics*. New York: Avon Books.

Mischel, W. (1970) Sex-Typing and Socialization. In L. Carmichael (ed.) *Manual of Child Psychology*. New York: Wiley.

Mitchell, J. (1973) *Woman's Estate*. New York: Vintage Books.

—— (1974) *Psychoanalysis and Feminism*. London: Allen Lane.

Möbius, J. P. (1901) The Physiological Mental Weakness of Woman. *Alienist and Neurologist* **22**: 624–42.

Montagu, A. (1937) The Origin of Subincision in Australia. *Oceania* **8**: 193–207.

Montgomery, R. E. (1974) A Cross-Cultural Study of Menstruation, Menstrual Taboos and Related Social Variables. *Ethos* **2**(2): 137–70.

Morris, C. (1979) Paradigms and Politics: The Ideas of the Women's Movement. *International Journal of Women's Studies* **2**(2): 189–201.

Morris, D. (1967) *The Naked Ape*. London: Jonathan Cape.

Mosedale, S. S. (1978) Science Corrupted: Victorian Biologists Consider 'The Woman Question'. *Journal of the History of Biology* **11**: 1–55.

Moss, P. (1976) The Current Situation. In N. Fonda and P. Moss (eds) *Mothers in Employment*. Uxbridge: Brunel University Management Programme.

Mosso, A. (1892) The Physical Education of Woman. *Pedagogical Seminary* **2**: 226–35.

Mulkay, M. (1979) *Science and the Sociology of Science*. London: Allen & Unwin.

Naismith, G. (1966) *Private and Personal*. New York: McKay.

National Organization of Women (1967) Bill of Rights. In R. Morgan (ed.) (1970) *Sisterhood is Powerful*. New York: Random House.

Nava, M. (1980) Gender and Education. *Feminist Review* **5**: 69–78.

New York Radical Women (1970) Principles. In R. Morgan (ed.) *Sisterhood is Powerful* New York: Random House.

Newcombe, F. and Ratcliff, G. (1973) Handedness, Speech Lateralization and Spatial Ability. *Neuropsychologia* **11**: 399–407.

Newcombe, F. and Ratcliff, G. (1978) The Female Brain: A Neuropsychological Viewpoint. In S. Ardener (ed.) *Defining Females*. London: Croom Helm.

Oakley, A. (1975) *The Sociology of Housework*. New York: Pantheon.

Ornstein, R. E. (1973) Right and Left Thinking. *Psychology Today* (May): 86–8, 90, 92.

Ortner, S. B. (1974) Is Female to Male as Nature is to Culture? In M. Z. Rosaldo and L. Lamphere (eds) *Woman, Culture and Society*. Stanford: Stanford University Press.

Osborn, S. M. and Harris, G. G. (1975) *Assertiveness Training for Women*. Springfield, Illinois: Thomas.

Page, M. (1978) Socialist Feminism – A Political Alternative? *m/f* **2**: 32–42.

Paige, K. E. (1973) Women Learn to Sing the Menstrual Blues. *Psychology Today* 7(4): 41–46.

Parchappe, M. (1848) Du Décroissement Graduel du Cerveau en Raison de la Dégradation Successive de l'Intelligence dans la Folie Simple. *Comptes Rendus Hebdomadaires des Séances de l'Académie des Sciences* **27**: 114–16.

Parke, R. and O'Leary, S. (1976) Mother-Father-Infant Interaction in the Newborn Period: Some Findings, Some Observations, and Some Unresolved Issues. In K. F. Riegel and J. A. Meacham (eds) *The Developing Individual in a Changing World*. Chicago: Aldine.

Parlee, M. B. (1973) The Premenstrual Syndrome. *Psychological Bulletin* **80**: 454–65.

—— (1978) The Sexes Under Scrutiny: From Old Biases to New Theories. *Psychology Today* **12**(6): 62–9.

Parsons, J. E., Ruble, D. N., Hodges, K. L., and Small, A. W. (1976) Cognitive-Developmental Factors in Emerging Sex Differences in Achievement-Related Expectancies. *Journal of Social Issues* **32**(3): 47–61.

Patrick, G. T. W. (1895) The Psychology of Woman. *Popular Science Monthly* **47**: 209–25.

Peacock, T. B. (1846) Tables of the Weights of Some of the Organs of the Human Body. *Monthly Journal of Medicine* **7**: 101–10, 166–78.

Pearson, K. (1902) On the Correlation of Intellectual Ability with the Size and Shape of the Head. *Royal Society Proceedings* **69**: 333–42.

Peck, E. and Senderowitz, J. (1974) Introduction. In E. Peck and J. Senderowitz (eds) *Pronatalism: The Myth of Mom and Apple Pie*. New York: Thomas Y. Crowell.

Peel, J. D. Y. (1972) Introduction. In J. D. Y. Peel (ed.) *Herbert Spencer on Social Evolution*. Chicago: University of Chicago Press.

Pembrey, M. S. (1913–14) Woman's Place in Nature. *Science Progress* **8**: 133–40.

Phelps, E. S. (1874) In J. W. Howe (ed.) *Sex and Education: A Reply to Dr E. H. Clarke's 'Sex in Education'*. Boston: Roberts Brothers.

Pilbeam, D. (1973) An Idea We Could Live Without: The Naked Ape. In A. Montagu (ed.) *Man and Aggression* New York: Oxford University Press.

Pizzey, E. (1974) *Scream Quietly or the Neighbours Will Hear*. Harmondsworth: Penguin.

Porter, C. (1980) *Alexandra Kollontai*. London: Virago.

Porteus, S. D. and Babcock, M. E. (1926) *Temperament and Race*. Boston: Gorham Press.

Quadagno, J. S. (1979) Paradigms in Evolutionary Theory: The Sociobiological Model of Natural Selection. *American Sociological Review* **44**: 100–09.

Redstockings (1969) Redstockings Manifesto. In R. Morgan (ed.) *Sisterhood is Powerful* (1970) New York: Random House.

Reed, E. (1970) Women: Caste, Class or Oppressed Sex? In *Problems of Women's Liberation*. New York: Pathfinder Press.

Rey, L. (1971) Comment on B. Magas, 'Sex Politics: Class Politics'. *New Left Review* **66**: 92–5.

Rich, A. (1977) *Of Woman Born*. New York: Bantam Books.

—— (1980) Compulsory Heterosexuality and Lesbian Existence. *Signs* **5**(4): 631–60.

Rodgers, J. (1977) Why So Few Female Geniuses? *Boston Herald American* (23rd February): 1, 4.

Rodman, H. (1915) Reported by G. MacAdam, in Feminist Apartment House to Solve Baby Problem. *New York Times* (24th January, part v): 9.

Roeder, B. *et al.* (1980) Radiation and Women's Rights. *Newsweek* (7th April): **21**.

Romanes, G. J. (1887) Mental Differences Between Men and Women. *Nineteenth Century* **21**: 654–72.

Romanes, G. J. (1893) *An Examination of Weismannism*. London: Longmans, Green & Co.

Rose, H. (1979) Hyper-Reflexivity – a New Danger for the Counter-Movements. In H. Nowotny & H. Rose (eds) *Counter-Movements in the Sciences*. Sociology of the Sciences Yearbook, Vol. III. Dordrecht, Holland: Reidel.

Rose, H. and Hanmer, J. (1976) Women's Liberation, Reproduction, and the Technological Fix. In D. L. Barker and S. Allen (eds) *Sexual Divisions and Society: Process and Change*. London: Tavistock.

Rosenberg, R. (1975) In Search of Woman's Nature, 1850–1920. *Feminist Studies* **3**: 141–54.

Rosenblatt, P. C. and Cunningham, M. R. (1976) Sex Differences in Cross-Cultural Perspective. In B. Lloyd and J. Archer (eds) *Exploring Sex Differences*. London: Academic Press.

Rossi, A. (1964) Equality Between the Sexes: An Immodest Proposal. *Daedalus* **93**: 607–52.

—— (1969) Sex Equality: The Beginnings of Ideology. In A. G. Kaplan and J. P. Bean (eds) *Beyond Sex-Role Stereotypes: Readings Towards a Psychology of Androgyny*. (1976) Boston: Little, Brown.

—— (1977) A Biosocial Perspective on Parenting. *Daedalus* **106**: 1–32.

Rousseau, J. J. (1911) *Emile; or, Education*. Translated by B. Foxley. London: J. M. Dent. Original edition, 1762.

Rowbotham, S. (1973) *Women's Consciousness, Man's World*. Harmondsworth: Penguin.

—— (1979) The Women's Movement and Organizing for Socialism. In S. Rowbotham, L. Segal and H. Wainwright, *Beyond the Fragments: Feminism and the Making of Socialism*. London: Merlin Press.

Rubin, G. (1975) The Traffic in Women: Notes on the 'Political Economy' of Sex. In R. Reiter (ed.) *Toward an Anthropology of Women*. New York: Monthly Review Press.

Ryan, J. (1972) IQ – The Illusion of Objectivity. In K. Richardson and D. Spears (eds) *Race, Culture, and Intelligence*. Harmondsworth: Penguin.

Saffioti, H. I. B. (1978) *Women in Class Society*. Translated by M. Vale. New York: Monthly Review Press.

Sahlins, M. (1976) *The Use and Abuse of Biology: An Anthropological Critique of Sociobiology*. Ann Arbor: University of Michigan Press.

Sayers, J. (1979) Anatomy is Destiny: Variations on a Theme. *Women's Studies International Quarterly* **2**: 19–32.

—— (1980a) Psychological Sex Differences. In Brighton Women and Science Group (eds) *Alice Through the Microscope*. London: Virago.

—— (1980b) Biological Determinism, Psychology and the Division of Labour by Sex. *International Journal of Women's Studies* **3**(3): 241–60.

Sayers, S. (1980) Forces of Production and Relations of Production in Socialist Society. *Radical Philosophy* **24**: 19–26.

Schaaffhausen, H. (1868) On the Primitive Form of the Human Skull. *Anthropological Review* **6**: 412–31.

Schultz, D. P. (1969) The Human Subject in Psychological Research. *Psychological Bulletin* **72**: 214–28.

Scott, H. (1979) Women in Eastern Europe. In J. Lipman-Blumen and J. Barnard (eds) *Sex Roles and Social Policy*. London: Sage.

Seaman, B. and Seaman, G. (1977) *Women and the Crisis in Sex Hormones*. New York: Rawson Associates.

Seccombe, W. (1973) The Housewife and her Labour under Capitalism. *New Left Review* **83**: 3–24.

Sedgwick, M. K. (1901) Some Scientific Aspects of the Woman Suffrage Question. *Gunton's Magazine* **20**: 333–44.

Sergi, G. (1892) *Per l'Educazione e la Coltura della Donna*. Reviewed by A. Mosso (1892).

Seward, G. (1944) Psychological Effects of the Menstrual Cycle on Women Workers. *Psychological Bulletin* **41**: 90–102.

Sharpe, S. (1976) *Just Like a Girl: How Girls Learn to be Women*. Harmondsworth: Penguin.

Sherman, J. (1977) Effects of Biological Factors on Sex-Related Differences in Mathematics Achievement. In L. H. Fox, E. Fennema, and J. Sherman, *Women and Mathematics: Research Perspectives for Change*. Washington: National Institute of Education.

—— (1978) *Sex-Related Cognitive Differences: An Essay on Theory and Evidence*. Springfield, Illinois: Charles C. Thomas.

—— (1979) Cognitive Performance as a Function of Sex and Handedness: An Evaluation of the Levy Hypothesis. *Psychology of Women Quarterly* **3**: 378–90.

Shields, S. A. (1975) Functionalism, Darwinism, and the Psychology of Women: A Study in Social Myth. *American Psychologist* **30**: 739–54.

Showalter, E. and Showalter, E. (1970) Victorian Women and Menstruation. *Victorian Studies* **14**: 83–9.

Silverman, M. G. (1977) Relations of Production, the Incest and Menstrual Tabus among Pre-Colonial Barbanans and Gilbertese. *Anthropologica* **19**(1): 79–97.

Simons, M. A. and Benjamin, J. (1979) Simone de Beauvoir: An Interview. *Feminist Studies* **5**(2): 330–45.

Slaughter, M. J. (1979) Feminism and Socialism. *Marxist Perspectives* **2**(3): 32–49.

Smith, A. (1937) *An Inquiry into the Nature and Causes of the Wealth of Nations*. New York: Random House. Original edition, 1776.

Smith, C. and Lloyd, B. (1978) Maternal Behaviour and Perceived Sex of Infant: Revisited. *Child Development* **49**: 1263–265.

Smith-Rosenberg, C. and Rosenberg, C. (1973) The Female Animal: Medical and Biological Views of Woman and her Role in Nineteenth-Century America. *Journal of American History* **60**: 332–56.

Snow, L. F. and Johnson, S. M. (1978) Myths about Menstruation: Victims of our own Folklore. *International Journal of Women's Studies* **1**(1): 64–72.

Solly, S. (1847) *The Human Brain*. London: Longman, Brown, Green and Longmans.

Solotaroff, H. (1898) On the Origin of the Family. *American Anthropologist* **11**: 229–42.

Spencer, H. (1852) A Theory of Population Deduced from the General Law of Animal Fertility. *Westminster Review* **57**: 468–501.

—— (1857) Progress: Its Law and Cause. In J. D. Y. Peel (ed.) (1972) *Herbert Spencer on Social Evolution*. Chicago: University of Chicago Press.

—— (1860) The Social Organism. In J. D. Y. Peel (ed.) (1972) *Herbert Spencer on Social Evolution*. Chicago: University of Chicago Press.

—— (1873) Psychology of the Sexes. *Popular Science Monthly* **4**: 30–8.

—— (1884) *Social Statics*. New York: Appleton. Original edition, 1850.

—— (1896) *The Principles of Biology*. New York: Appleton. Original edition, 1867.

—— (1898) *The Principles of Sociology*, Vol. 1. New York: Appleton. Original edition, 1876.

Star, S. L. (1979) The Politics of Right and Left: Sex Differences in Hemispheric Brain Asymmetry. In R. Hubbard, M. S. Henifin, and B. Fried (eds) *Women Look at Biology Looking at Women*. Boston: G. K. Hall.

Stephens, W. N. (1961) A Cross-Cultural Study of Menstrual Taboos. *Genetic Psychology Monographs* **64**: 385–416.

Sternglanz, S. H. and Serbin, L. A. (1974) Sex Role Stereotyping in Children's Television Programmes. *Developmental Psychology* **10**: 710–15.

Stevenson, S. H. (1881) *The Physiology of Woman*. Chicago: Cushing, Thomas & Co.

Stoller, R. (1976) Primary Femininity. *Journal of the American Psychoanalytic Association* **24**(5): 59–78.

Stone, B. (1972) Women and Political Power. In L. Jenness (ed.) *Feminism and Socialism*. New York: Pathfinder Press.

Storer, H. R. (1868) *Criminal Abortion*. Boston: Little, Brown.

Sutherland, A. (1900) Woman's Brain. *Nineteenth Century* **47**: 802–10.

Swinburne, J. (1902) Feminine Mind Worship. *Westminster Review* **158**: 187–98.

Tanner, N. and Zihlman, A. (1976) Women in Evolution. Part I: Innovation and Selection in Human Origins. *Signs* **1**(3): 585–608.

Thomas, M. C. (1908) Present Tendencies in Women's College and University Education. *Educational Review* **35**: 64–85.

Thomas, W. I. (1897) On a Difference in the Metabolism of the Sexes. *American Journal of Sociology* **3**: 31–63.

—— (1898) The Relation of Sex to Primitive Social Control. *American Journal of Sociology* **3**: 754–76.

Thompson, C. (1943) 'Penis Envy' in Women. In J. B. Miller (ed.) *Psychoanalysis and Women*. Harmondsworth: Penguin. 1973.

Thompson, H. B. (1903) *The Mental Traits of Sex*. Chicago: University of Chicago Press.

Thompson, M. (1914) A Feminist Symposium. *New Review* (August): 444–47. Cited in M. J. Slaughter (1979).

Thorburn, J. (1884) *Female Education from a Physiological Point of View*. Manchester: J. E. Cornish.

Tiedmann, F. (1836) On the Brain of the Negro, Compared with that of the European and the Orang-Utang. *Philosophical Transactions* **126**: 497–527.

Tieger, T. (1980) On the Biological Basis of Sex Differences in Aggression. *Child Development* **51**: 943–63.

Tiger, L. (1970) The Possible Biological Origins of Sexual Discrimination. *Impact of Science on Society* **20**: 29–44.

Tiger, L. and Fox, R. (1974) *The Imperial Animal*. New York: Dell.

Tilt, E. J. (1874) President's Address, delivered at the Obstetrical Society on 6th January. Reported in the *British Medical Journal* (1875) **1**: 73.

Timpanaro, S. (1975) *On Materialism*. Translated by L. Garner. London: New Left Books.

Tomes, N. (1978) A 'Torrent of Abuse': Crimes of Violence Between Working-Class Men and Women in London, 1840–1875. *Journal of Social History* **11**: 328–45.

Topinard, P. (1885) *Éléments d'Anthropologie Générale*. Paris: Delahaye et Lecrosnier.

—— (1894) *Anthropology*. Translated by R. T. H. Bartley. London: Chapman & Hall. Original edition, 1876.

Trivers, R. L. (1971) The Evolution of Reciprocal Altruism. *Quarterly Review of Biology* **46**: 35–57.

—— (1972) Parental Investment and Sexual Selection. In B. Campbell (ed.) *Sexual Selection and the Descent of Man 1871–1971*. Chicago: Aldine.

Turkle, S. (1980) French Anti-Psychiatry. In D. Ingleby (ed.) *Critical Psychiatry*. New York: Pantheon.

U.S. Senate (1972) Debate on 'Equal Rights for Men and Women'. *Congressional Record* **118**(8): 9517–23.

Ullian, D. Z. (1976) The Development of Conceptions of Masculinity and Femininity. In B. B. Lloyd and J. Archer (eds) *Exploring Sex Differences*. London: Academic Press.

Van de Warker, E. (1875) Sexual Cerebration. *Popular Science Monthly* **7**: 289–92.

—— (1906) The Fetish of the Ovary. *American Journal of Obstetrics* **54**(3): 371.

Veblen, T. (1899) The Barbarian Status of Women. *American Journal of Sociology* **4**: 503–14.

Verral, R. (1979) *Spearhead* (March). Cited in The Sociobiology Study Group of the Boston Chapter (1979) A New Wave of Reaction in Europe: Sociobiology Used to Justify Racism, Sexism and Elitism. *Science for the People* **11**(6): 29.

Vogt, K. (1864) *Lectures on Man*. London: Longman.

Von Soemmerring, S. T. (1788) *S.Th. Soemmerring vom Hirn und Rückenmark*. Mainz: P. A. Winkopp. Cited in E. Fee (1979).

Waber, D. P. (1976) Sex Differences in Cognition: A Function of Maturation Rates? *Science* **192**: 572–74.

Walker, A. (1840) *Woman Physiologically Considered*. New York: J. & H. G. Langley.

Walsh, M. R. (1977) *'Doctors Wanted: No Women Need Apply': Sexual Barriers in the Medical Profession, 1835–1975*. New Haven: Yale University Press.

Washburn, S. L. and DeVore, I. (1961) The Social Life of Baboons. *Scientific American* **204**(6): 62–71.

Wedge, P. and Prosser, H. (1973) *Born to Fail?* London: Arrow Books.

Weideger, P. (1975) *Menstruation and Menopause*. New York: Knopf.

Weir, J. (1895) The Effect of Female Suffrage on Posterity. *American Naturalist* **29**: 815–25.

Weitzman, L. J., Eifler, D., Hokada, E., and Ross, C. (1972) Sex Role Socialization in Picture Books for Preschool Children. *American Journal of Sociology* **77**: 1125–250.

Whitehead, R. E. (1934) Notes from the Department of Commerce: Women Pilots. *Journal of Aviation Medicine* **5**: 47–9.

Wilson, Edward O. (1975a) *Sociobiology: The New Synthesis*. Cambridge, Mass.: Harvard University Press.

—— (1975b) Human Decency is Animal. *New York Times Magazine* (12th October): 38–40, 42–6, 48, 50.

— (1978) *On Human Nature*. Cambridge: Harvard University Press.

Wilson, Elizabeth (1980) Beyond the Ghetto: Thoughts on 'Beyond the Fragments – Feminism and the Making of Socialism' by Hilary Wainwright, Sheila Rowbotham and Lynne Segal. *Feminist Review* **4**: 28–44.

Winnicott, D. W. (1960) The Theory of the Parent-Infant Relationship. *International Journal of Psychoanalysis* **41**: 585–95.

Witelson, S. F. (1976) Sex and the Single Hemisphere: Specialization of the Right Hemisphere for Spatial Processing. *Science* **193**: 425–27.

Wollstonecraft, M. (1792) *A Vindication of the Rights of Woman*. London: J. Johnson.

Woody, T. (1929) *A History of Women's Education in the United States*, Vol. II. New York: The Science Press.

Woolley, H. T. (1910) Psychological Literature. A Review of the Recent Literature on the Psychology of Sex. *Psychological Bulletin* **7**: 335–42.

Wright, A. (1912) Letter to *The Times* (28th March): 7–8.

Yates, G. G. (1975) *What Women Want: The Ideas of the Movement*. Cambridge: Harvard University Press.

Young, R. M. (1970) *Mind, Brain and Adaptation in the Nineteenth Century*. Oxford: Clarendon Press.

—— (1977) Science *is* Social Relations. *Radical Science Journal* **5**: 65–131.

—— (1980) The Relevance of Bernal's Questions. *Radical Science Journal* **10**: 85–94.

Zaretsky, E. (1976) *Capitalism, the Family, and Personal Life*. New York: Harper & Row.

—— (1979) A Note on Marxism and Psychoanalysis. *Marxist Perspectives* **2**(2): 134–40.

Zelman, E. C. (1977) Reproduction, Ritual, and Power. *American Ethnologist* **4**: 714–33.

Zetkin, C. (1896) Only with the Proletarian Woman will Socialism be Victorious. Reprinted in H. Draper and A. G. Lipow (1976).

Name index*

*Thanks to Ian Toulson for his computer-assistance in compiling this index.

Subject index